Tannins - Structural Properties, Biological Properties and Current Knowledge

Edited by Alfredo Aires

Published in London, United Kingdom

IntechOpen

Supporting open minds since 2005

Tannins - Structural Properties, Biological Properties and Current Knowledge
http://dx.doi.org/10.5772/intechopen.80170
Edited by Alfredo Aires

Contributors
Dang Xuan Cuong, Dang Xuan Chinh, Nguyen Xuan Hoan, Dinh Huu Dong, Nguyen Van Thanh, Le Thi Minh Thuy, Hoang Thai Ha, Dang Thi Thanh Tuyen, Md Nazrul Islam, Atanu Kumar Das, Md. Omar Faruk, Md. Ashaduzzaman, Rudi Dungani, Enih Rosamah, Sri Hartati, Alfi Rumidatul, Sergio Miele, Stefania Tegli, Enrica Bargiacchi, Matteo Cerboneschi, Carlos Garcia Izquierdo, Xiaojian Zhou, Guanben Du, Angel S. Galabov, Neli Vilhelmova-Ilieva, Milka Mileva, Akhlash P. Singh, Sunil Kumar, Fabrice Vaillant Barka, Karen Johana Ortega Villalba, Carlos Vélez Pasos, Pablo Emilio Rodríguez

First published in London, United Kingdom, 2020 by IntechOpen
IntechOpen is the global imprint of INTECHOPEN LIMITED, registered in England and Wales, registration number: 11086078, 7th floor, 10 Lower Thames Street, London, EC3R 6AF, United Kingdom
Printed in Croatia

British Library Cataloguing-in-Publication Data
A catalogue record for this book is available from the British Library

Additional hard and PDF copies can be obtained from orders@intechopen.com

Tannins - Structural Properties, Biological Properties and Current Knowledge
Edited by Alfredo Aires
p. cm.
Print ISBN 978-1-78984-796-3
Online ISBN 978-1-78984-797-0
eBook (PDF) ISBN 978-1-83968-358-9

For EU product safety concerns:
IN TECH d.o.o., Prolaz Marije Krucifikse Kozulić 3, 51000 Rijeka, Croatia, info@intechopen.com or visit our website at intechopen.com.

Meet the editor

Alfredo Aires (PhD in Agronomic Sciences from the University of Trás-os-Montes e Alto Douro, Portugal) was born on March 23,1973 in Vila Real, northern Portugal. He is an expert in the area of agronomy and food composition analysis with special emphasis on secondary plant metabolites, their variation, and health effects. He has produced 116 publications: 40 papers published in journals indexed in the Scientific Citation Index, six chapters in books, and several other papers in peer-reviewed journals in abstract books. His H-factor is 14 by the Web of Knowledge from Thomson Reuters. Currently, he is a researcher at the Centre for the Research and Technology of Agro-Environmental and Biological Sciences, CITAB, at the University of Trás-os-Montes and Alto Douro, Vila Real, Portugal. His interests are extraction and isolation of secondary metabolites, including polyphenols and tannins, with importance in agriculture and the food by-products industry.

Contents

Preface

Tannins are one of the polyphenols group found in plants and are mainly studied because of their structural properties and bioactive behavior. Every year new findings concerning their properties and functions are made, and today concerns are mainly focused on how they can be used efficiently in the wood, food, textile, health, and pharmaceutical industries. Thus, the aim of this book is to present the most updated information on the structural properties of tannins, their food sources and variations, biological properties, and health, among other important issues. In addition, the most recent methods used for their isolation, quantifications, and industrial applications will also be covered.

Alfredo Aires (PhD)
CITAB—Centre for the Research and Technology of Agro-Environmental and Biological Sciences,
University of Trás-os-Montes and Alto Douro,
Agronomy Department,
Vila Real, Portugal

Section 1

Sources, Utilization and Biological Activity

Chapter 1

Hardwood Tannin: Sources, Utilizations, and Prospects

Atanu Kumar Das, Md. Nazrul Islam, Md. Omar Faruk, Md. Ashaduzzaman, Rudi Dungani, Enih Rosamah, Sri Hartati and Alfi Rumidatul

Abstract

Tannins are found in widely distributed species of plants, and it protects plant from predators and pests. There are three major groups of tannins, that is, hydrolyzable, complex, and proanthocyanidins. Tannins are being used as a significant element for the tanning of animal hides in the leather production industry from the beginning of tannin industry. Then, these have been used for mineral absorption and protein precipitation purposes since the 1960s. Tannins are used for iron gall ink production and wood-based industry as adhesive and anticorrosive, recovering uranium from seawater and removing mercury and methylmercury from solution. In addition, tannins are considered as bioactive compound in nutrition science, and their possible effects on health are to be identified. This chapter outlines the structural and biological properties of hardwood tannins to indicate the positive utilization of them. It also describes the contemporary information on tannins.

Keywords: tannin, structural properties, biological properties, utilization

1. Tannin and its classification

Plants are protected from herbivores and diseases due to the accumulation of a wide range of "secondary" compounds, together with alkaloids, terpenes, and phenolics. Phenolic metabolism helps to produce a wide variety of compounds ranging from the familiar anthocyanidins (flower pigments) to lignin (complex phenolics of the plant cell wall). Generally, the chemical and biological criteria of phenolic compounds are called as tannins which are different from other plant secondary phenolics. The word "tannin" originates from the ancient Celtic word for oak [1].

Seguin has introduced the term "tannin" to explain the ability to convert hide or skin into leather by organic material extracted from certain plant tissues in 1796 [1]. Tannin is a chemical component obtained from plant, and it has been used for the benefit of human being from the beginning of its development. To get the idea about its complete nature, a lot of research work is needed till today, and that is why it is considered the "dark continents" of science. Tannin has the ability to make proteins and other polymers like pectin [2].

Tannins are the main polyphenolic secondary metabolites distributed widely in the range of 5–10% of dry plant material in vascular plants. They are found mainly in the bark, stems, seeds, roots, buds, and leaves [3]. Tannins are either galloyl

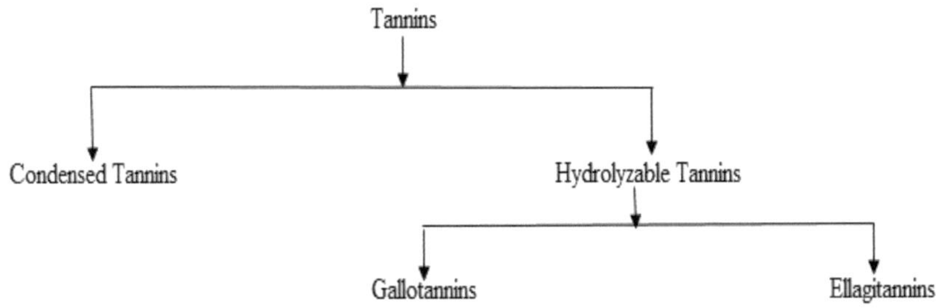

Figure 1.
Classification of tannins.

esters, or they are oligomeric and polymeric proanthocyanidins. They protect trees from fungi, pathogens, insects, and herbivorous animals [4].

Tannins in softwood and hardwood are different as softwood and hardwood dust particles can be separated based on their tannin content [5].

The term "tannin" is difficult to define concisely as it is found in plant chemistry. Generally, tannin contains precise physical and chemical components, which help to convert hide or skin of animal into impermeable non-rotting leather. Broadly, the definition of tannin covers a whole mass of components which give overall phenolic reactions [2].

Bate-Smith has defined tannins in this way: "tannins are water soluble phenolic compounds with molecular weight of 500 and 3000, capable of precipitating gelatin, alkaloids and other proteins and provide the typical phenolic reactions" [1, 6]. Tannins having a high molecular weight of 20,000 have shown complex with some specific polysaccharides [1, 6]. Scientists have also used the term "polyphenol" in place of "tannin" to emphasize the diversity of phenol groups which characterize these compounds [7, 8].

Natural tannins are generally categorized as either condensed or hydrolyzable tannins (**Figure 1**). Hydrolyzable tannins are subdivided into gallotannins and ellagitannins [4].

Condensed tannins have a wide range of molecular weight from 500 to over 20,000 [9]. Condensed tannins are composed of flavolans or polymers of flavan-3-ols (catechins) and/or flavan 3:4-diols (leucoanthocyanidins) [10]. Phlobaphene, a water insoluble product, is possible to be produced by polymerization of condensed tannins. Condensed tannins have the ability to react with aldehydes to generate polymeric materials [9]. The hydrolyzable tannins contain glucose or other polyhydric alcohols esterified with gallic acid (gallotannins) or hexahydroxydiphenic acid (ellagitannins) [11]. Acid hydrolysis of hydrolyzable tannins helps to get gallic acid, ellagic acid, or other similar species [7]. The most important condensed tanning materials are wattle or "mimosa," quebracho, mangrove, and hemlock, whereas the hydrolyzable tanning materials are chestnut wood (*Castanea sativa* and *C. dentata*) and dry myrobalan fruits (*Terminalia chebula*) [10].

2. Hardwood tannin

2.1 Condensed tannin

Condensed tannins are polymeric flavonoids, and most of them are based on the flavan-3-ols (**Figure 2**) (−)-epicatechin and (+)-catechin [1].

The best categorized condensed tannins are linked by C8 of the terminal unit and C4 of the extender. The four common modes of coupling are called B-1, B-2,

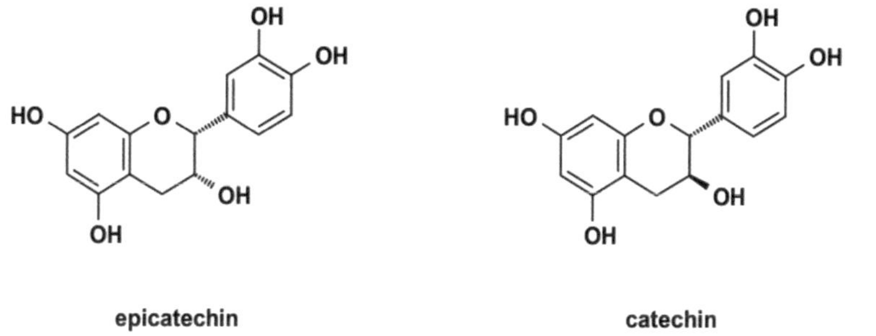

Figure 2.
Flavan-3-ols [1].

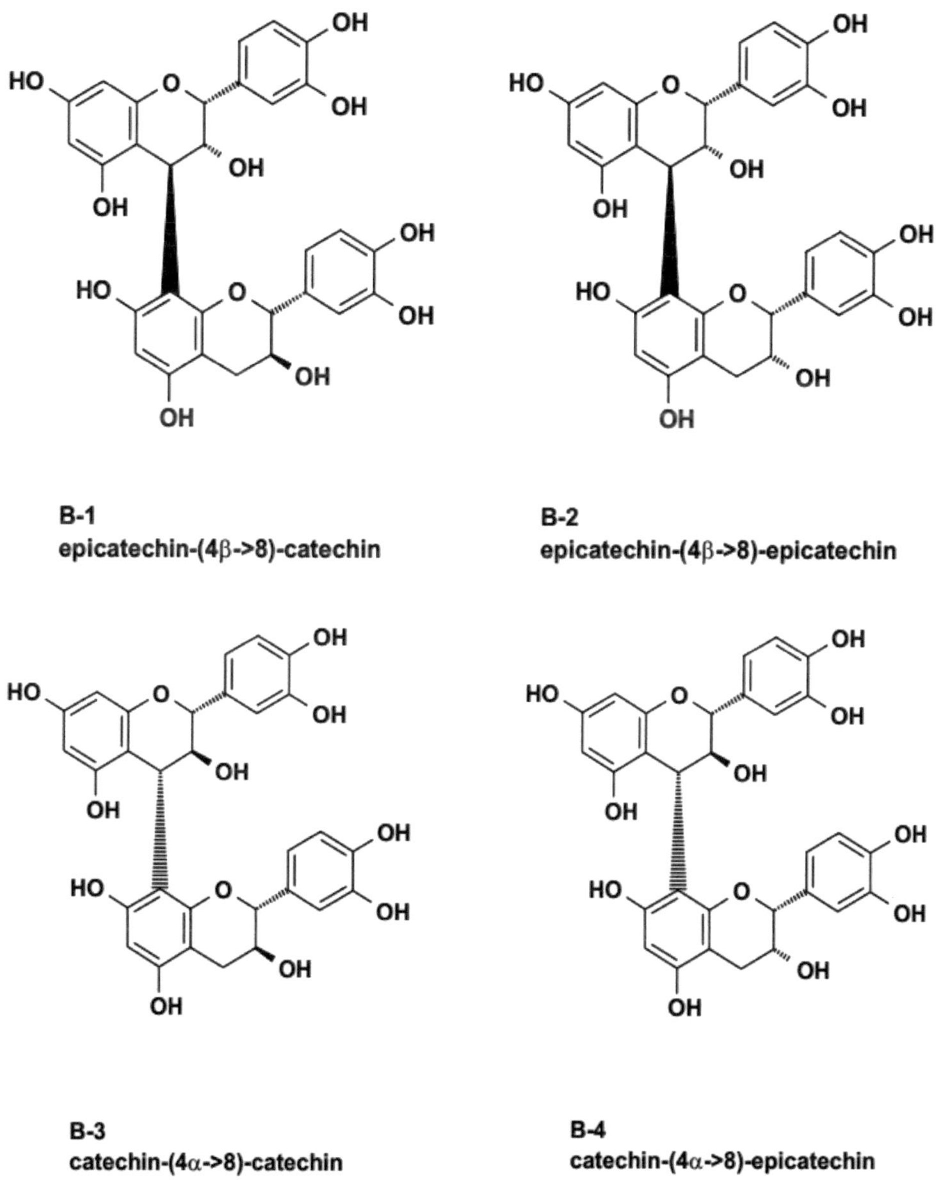

Figure 3.
Common modes of coupling of condensed tannins [1].

B-3, and B-4 (**Figure 3**). Besides these dimers, related dimers which are linked by C6 of the terminal unit and C4 of the extender have been separated [1].

Proanthocyanidin is becoming more popular to describe flavonoid-based polyphenolics instead of condensed tannins. Anthocyanidin pigments are derived from proanthocyanidins by oxidative cleavage in hot alcohols (**Figure 4**) [1].

Unmodified terminal units are produced by the acid butanol reaction, and the extender units produce the colored anthocyanidins. Procyanidins are produced by catechin- and epicatechin-based polymers. Delphinidin is the yield of gallocatechin- and epigallocatechin-based polymers, whereas pelargonidin is the product of the rare mono-substituted flavan-3-ol-based polymers.

There is an available branch in 5-deoxy-flavan-3-ol polymer-condensed tannins due to having the reactivity of the 5-deoxy A ring, and these are an important group of condensed tannins. In the case of quebracho and acacia tannin preparations, profisetinidins and prorobinetinidins are observed as the major tannins. Acid butanol reaction helps to produce the 5-deoxy anthocyanidins fisetinidin and robinetinidin (**Figure 5**) [1].

Tannins are broadly available in all over the plant kingdom, while the condensed tannins are distributed in the Archichlamydeae division of the

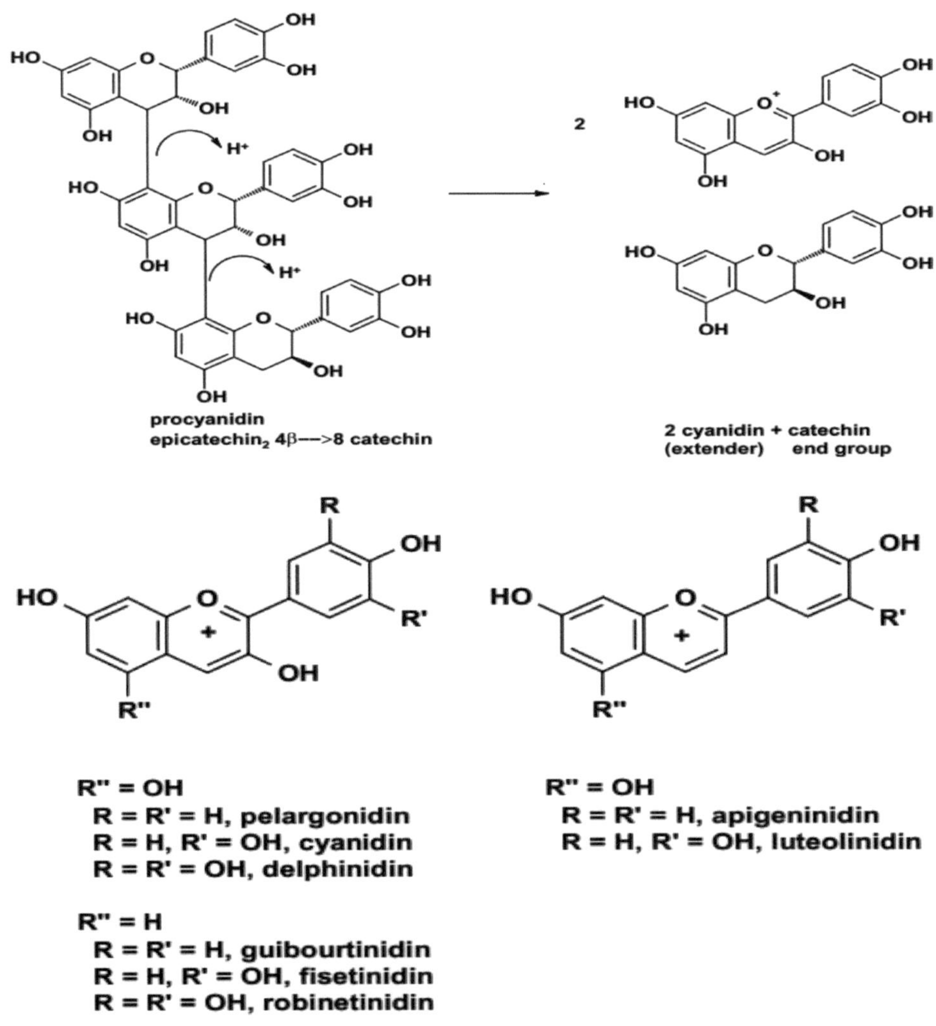

Figure 4.
Proanthocyanidin and its derivative [1].

DOI: http://dx.doi.org/10.5772/intechopen.86003

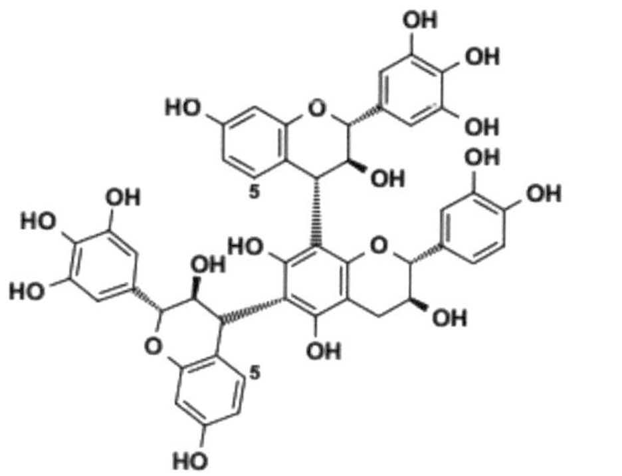

robinetinidol-(4α->8)-catechin-(6α->4a)-robinetinidol

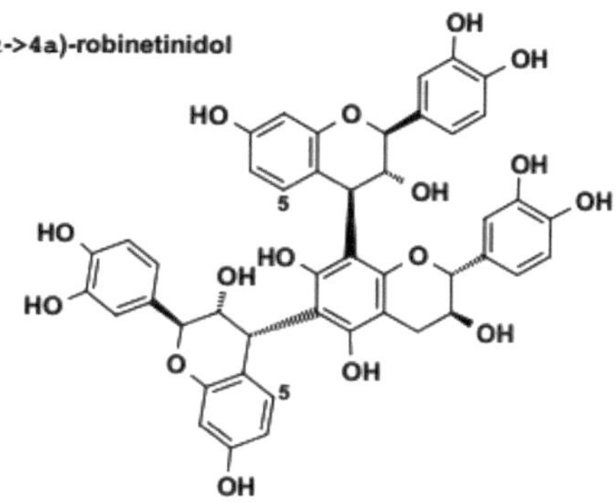

profisetinidin

Figure 5.
Extracted tannins by acid butanol reaction [1].

Family	Genus
Casuarinaceae	*Casuarina*
Salicaceae	*Salix*
Betulaceae	*Betula*
Fagaceae	*Castanea*, *Lithocarpus*, *Quercus*
Lauraceae	*Persea*
Rosaceae	*Prunus*
Leguminosae	*Acacia*, *Cassia*, *Robinia*
Meliaceae	*Carapa*, *Xylocarpus*
Euphorbiaceae	*Excoecaria*
Anacardiaceae	*Astronium*, *Rhus*, *Schinopsis*
Bombacaceae	*Camptostemon*
Sterculiaceae	*Heritiera*
Dipterocarpaceae	*Hopea*
Lythraceae	*Pemphis*
Sonneratiaceae	*Sonneratia*
Rhizophoraceae	*Bruguiera*, *Carallia*, *Ceriops*, *Candelaria*, *Rhizophora*
Combretaceae	*Laguncularia*
Myrtaceae	*Eucalyptus*, *Tristania*
Verbenaceae	*Avicennia*

Table 1.
Condensed tannins in hardwood.

Dicotyledonae. Scientists examined condensed tannins by paper chromatography in their previous studies [12–16]. In Australia, the attractive source of condensed tannins are extracts from barks of *Bruguiera* and *Rhizophora* species (mangrove species) and *Eucalyptus astringens* (brown mallet) and the heartwood and bark of *E. wandoo* and *E. accedens* (wandoos). **Table 1** shows the distribution of condensed tannins in hardwood.

2.1.1 Utilization of condensed tannin

2.1.1.1 Tanning industry

In general, condensed tannin material is the major component of commercial tanning chemical, which is used for tanning leather in tannery. The major contributors for the world's supply of condensed tannin are extract of mangrove and wattle and quebracho used in heavy leather manufacture. Condensed tannin materials are selected based on their tanning quality for heavy leather production. Condensed tannins have high resistance power to detanning [17]. Quebracho and wattle tannins are more resistant to concentrated urea solution than to hydrolyzable tannins [18].

Though tannin materials are dissimilar based on pH, salt content, and natural conditions of acid content, it can be controlled maintaining proper conditions [19, 20]. pH in tan liquors is relatively high due to the absence of phenolic in tan liquors. Mangrove extracts contain sodium chloride, whereas sulfited quebracho extracts possess sodium sulfate, and the salt content contributes to the salt material in tannin materials [21].

2.1.1.2 Preservative for fishing net

Preservation of fishing nets by various condensed tannin materials is popular in Indo-Pacific countries. It is used to prevent cellulose degradation by bacterial and fungal cellulases. Fishing nets are submerged in hot tannin solution, and this method is repeated for several times. Tannin-impregnated nets are further treated with either hot ammoniacal copper sulfate or dichromate solution for some cases [22, 23]. Condensed tannin materials are extracted from Burma cutch from *Acacia catechu* heartwood, Malayan (mangrove) cutch from *Carapa obovata* and *Ceriops candolleana* barks, and Borneo (mangrove) cutch from *Ceriops candolleana*, *Rhizophora candelaria*, and *R. mucronata* barks [21].

2.1.1.3 Preparation of plastics and adhesives

Condensed tannins are phenolic raw materials which react with formaldehyde; these can be used for the production of synthetic resins [21]. Nico [24] studied the suitability of quebracho-formaldehyde resins for adhesives and plastics. Molding powders are produced by pressing mixtures of accurate quebracho and wattle tannins, paraformaldehyde, and plasticizer [21]. Though these moldings have short flow period, these follow all the characteristics of phenol-formaldehyde moldings as per the specifications of British standards [25]. These molding powders show very stable and waterproof properties, and the mixture is catalyzed with either acid or alkali at temperature 250°F [25].

Various condensed tannins as plywood and particle board adhesives have been studied by many researchers [26–30]. Adhesives prepared from bark of *Tsuga heterophylla*, *Acacia mollissima*, and *Callitris calcarata* have less strength and higher water resistance properties than those obtained from commercial adhesives [31, 32]. Strong and water-resistant adhesives for plywood are prepared by mixing of paraformaldehyde, filler, and sulfited tannin solutions obtained from *Eucalyptus crebra*, *Pinus radiata*, wattle extract, and quebracho extract [25, 33]. Adhesives prepared from Australian mangrove tannin extracts in addition with a small proportion of commercial phenol-formaldehyde, resorcinol-formaldehyde, or phenol- and resorcinol formaldehyde comply with Australian and British standard for synthetic adhesives for plywood [21, 34].

2.1.1.4 Oil and ceramic industry

The main problem is to maintain the flow and suspension characteristics of bentonite oil well muds in the United States [21]. In oil well drilling, quebracho tannins are used widely to control the viscosity of the mud; annual consumption of quebracho extract is 30,000–40,000 tons for this purpose [35]. In the United States, it is accounted for about 40% of the total tannin consumption in 1950 [36, 37].

Scientist has reviewed the incorporation of condensed tannins into the ceramic industry. Quebracho extract enables the use of higher solid mixer in the casting of clay slip by lowering the viscosity of the clay-water mixes. Condensed tannins save the plaster molds from deterioration by increasing the tensile strength of clay casts and eliminating the silica from the mixture, with consequent saving in the deterioration of plaster molds. The suspending power of the slip is enhanced by quebracho tannins in the casting of bone China [21].

2.1.1.5 Anticorrosive of metals

Tannins protect the iron materials from sulfate-reducing bacteria by exerting a bacteriostatic action [38]. Mangrove tannin is economically viable to protect the underground iron pipes and tubes [35].

Tannate films of tannin extract is formed on iron and steel surfaces to protect it from atmospheric corrosion [25]. Actually a protective film is formed during the submersion of aluminum in tannin solutions, and it can be used for the preservation of nonferrous metal [39].

2.2 Hydrolyzable tannin

Hydrolyzable tannins are esters of a sugar with one or more polyphenolic carboxylic acids; sugar is mainly glucose, but polysaccharides or branched-chain sugars are available for some cases. The hydrolyzable tannins are generally classified based on the hydrolyzed product, and these are either gallotannins, are simple polygalloyl esters of glucose. Pentagalloyl glucose, has five identical ester linkages that involve aliphatic hydroxyl groups of the core sugar (**Figure 6**). Alkalies, acids, or enzymes like tannase and takadiastase hydrolyze the ester linkages in hydrolyzed tannins [40].

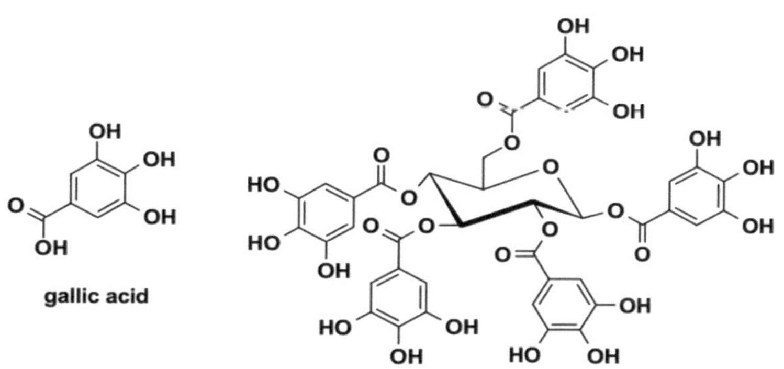

Figure 6.
Precursors of gallotannins [1].

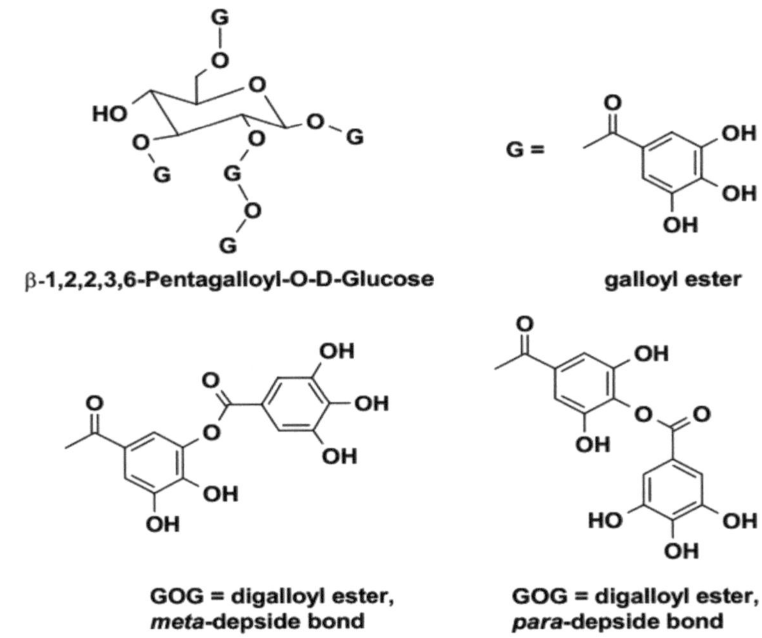

Figure 7.
Isomers of PGG [1].

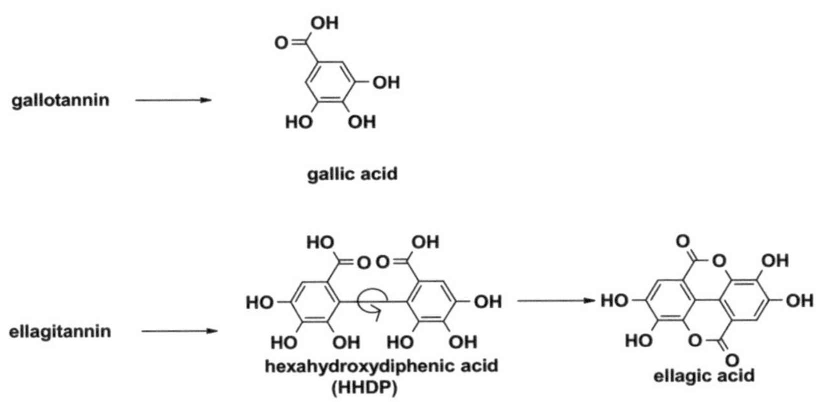

Figure 8.
Ellagitannins [1].

Gallotannins, the simplest hydrolyzable tannins, are simple polygalloyl esters of glucose. The ideal gallotannin is comprised esters of β-1,2,3,4,6-pentagalloyl-*O*-D-glucopyranose (pentagalloylglucose) (**Figure 6**). Pentagalloylglucose (PGG) has five identical ester linkages having aliphatic hydroxyl groups of the core sugar [21].

PGG has many isomers having the same molecular weights (940 g/mol) (**Figure 7**). Although the molecular weights are same, chemical properties and biochemical properties are different from each other due to having different structures. Therefore, their properties like susceptibility to hydrolysis, chromatographic behavior, and ability to precipitate protein are varied for different isomers [21].

Gallotannins are converted to the related ellagitannins by the oxidative coupling of galloyl groups (**Figure 8**). The simple ellagitannins are esters of hexahydroxydiphenic acid (HHDP). In aqueous solution, HHDP impulsively lactonizes to ellagic acid [21].

HHDP is formed by intramolecular carbon-carbon coupling C-4/C-6 (e.g., eugeniin) and C-2/C-3 (e.g., casuarictin, also has C-4/C-6) having polygalloyl glucoses

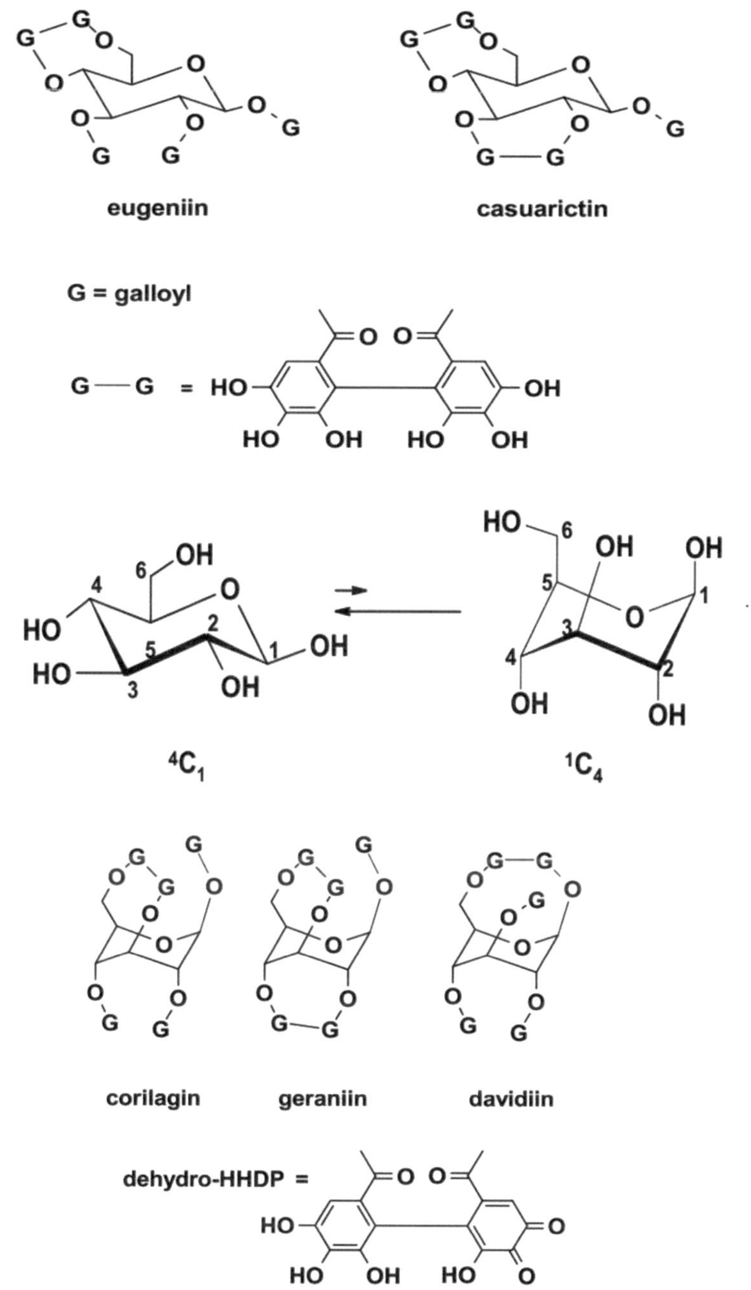

Figure 9.
Hexahydroxydiphenic acid (HHDP) [1].

in the more stable $^4C_1$ conformation (**Figure 9**). On the other hand, intramolecular coupling occurs at C-3/C-6 (e.g., corilagin), C-2/C-4 (e.g., geraniin, also has C-3/C-6), or C-1/C-6 (e.g., davidiin) with polygalloyl glucose the less stable $^1C_4$ conformation for a few plants (**Figure 9**). In addition, geraniin is characterized by partial oxidation of the C-2/C-4 HHDP to dehydro-HHDP [21].

Some commercial hydrolyzable tannins presented in **Table 2** develop the present knowledge on the chemistry of the hydrolyzable tannins. Hydrolyzable tannins are available in the leaves and sapwood of a tree [40, 41]. Ellagitannins (or ellagic acid) are found in the leaves of species of the orders *Fagales*, *Myrtiflorae*, *Rosales*,

Tannin	Source
Chinese tannin (tannic acid)	Galls on leaves of *Rhus semialata* (Anacardiaceae)
Sumac tannin	Leaves of *Rhus coriaria* (Anacardiaceae)
Myrobalans	Fruit of *Terminalia chebula* (Combretaceae)
Turkish tannin	Galls on wood of *Quercus infectoria* (Fagaceae)
Valonea extract	Acorn cups of *Quercus valonea* or *Q. macrolepis* (Fagaceae)
Oak extract	Wood of various *Quercus* species (Fagaceae)
Chestnut extract	Wood of *Castanea sativa* and *C. dentata* (Fagaceae)
Tara extract	Pods of *Caesalpinia spinosa* (Leguminosae)
Divi-divi	Pods of *Caesalpinia coriaria* (Leguminosae)
Algarobilla extract	Pods of *Caesalpinia brevifolia* (Leguminosae)

Table 2.
Name of commercial hydrolyzable tannins and sources.

Family	Genus
Casuarinaceae	*Casuarina*
Juglandaceae	*Pterocarya*
Betulaceae	*Alnus*, *Carpinus*
Fagaceae	*Castanea*, *Nothofagus*, *Quercus*
Loranthaceae	*Nuytsia*
Nymphaeaceae	*Nuphar*
Cercidiphyllaceae	*Cercidiphyllum*
Droseraceae	*Drosera*
Saxifragaceae	*Francoa*, *Ribes*
Hamamelidaceae	*Liquidambar*
Rosaceae	Rosoideae (*Geum*, *Fragaria*, *Potentilla*, *Rosa*, *Rubus*) only
Leguminosae	*Pterocarpus* (*Caesalpinia* in lit.)
Geraniaceae	*Geranium*
Tremandraceae	*Tetratheca*
Euphorbiaceae	*Acalypha*, *Euphorbia*, *Ricinus*
Empetraceae	*Empetrum*
Coriariaceae	*Coriaria*
Cyrillaceae	*Cyrilla*
Corynocarpaceae	*Corynocarpus*
Aceraceae	*Acer monspessulanum* (*A. ginnala* in lit.)
Melianthaceae (Anacardiaceae in lit.)	*Greyia*, *Melianthus*
Vitaceae	*Vitis*
Elaeocarpaceae	*Aristotelia*
Theaceae	*Camellia*, *Cleyara*, *Gordonia*, *Thea*
Frankeniaceae	*Frankenia*
Tamaricaceae	*Tamarix*
Bixaceae	*Bixa*
Stachyuraceae	*Stachyurus*
Elaeagnaceae	*Elaeagnus*, *Hippophae*

Family	Genus
Lythraceae	*Lagerstroemia*, *Lythrum*
Punicaceae	*Punica*
Rhizophoraceae	*Cassipourea* (*Rhizophora* in lit.)
Combretaceae	*Combretum* (*Terminalia* in lit.)
Myrtaceae	*Agonis*, *Angophora*, *Callistemon*, *Eucalyptus*, *Eugenia*, *Melaleuca*, *Metrosideros*, *Myrtus*, *Psidium*
Melastomataceae	*Bertolonia*, *Heterocentron*, *Medinilla*, *Tibouchina*
Onagraceae	*Dircaea*, *Fuchsia*, *Jussieua*, *Lopezia*, *Oenothera*
Haloragaceae	*Gunnera*, *Haloragis*, *Myriophyllum*
Ericaceae	*Arbutus* (*Arctostaphylos*, *Vaccinium*, in lit.)
Diapensiaceae	*Galax*
Campanulaceae	*Centropogon* only
Compositae	*Tagetes* only
Amaryllidaceae	*Hypoxis*

Table 3.
Name of family and genus containing ellagitannins or ellagic acid.

Sapindales, and *Geraniales* [42]. Gallotannins and ellagitannins have been detected in the tissues of many hundreds of species of dicotyledons [40]. Source of ellagitannins or ellagic acid is presented in **Table 3**.

2.2.1 Utilization of hydrolyzable tannin

2.2.1.1 Tanning industry

For leather manufacturing process, the conversion of animal hide or skin into leather is considered as the main art of tanning in leather industries. To provide leather, tannin cross-links with the collagen chains located in the hide during tanning process. Hydrolyzable tannins obtained from plant extract are widely used for tanning in leather industries [2].

2.2.1.2 Medication

Biological activities of hydrolyzable tannins confirm the beneficial effect on the health of human being, and these are used as antimutagenic, anticancer, and antioxidant. Furthermore, hydrolyzable tannins help to diminish serum cholesterol and triglycerides and suppress lipogenesis by insulin [43, 44].

2.2.1.3 Other applications

Tannic acid is used in dye industry and water treatment process for purifying water. It is also used for manufacturing ink, plastic resins, adhesives, surface coatings, gallic acid, etc. [2].

2.2.2 Prospectus of the utilization of tannins

There are several uses of tannins for the welfare of human being, and continuous research is being carried out for increasing the diversity of utilization of tannins. The potential uses of tannins are in the following.

2.2.2.1 Tanning industry

Tannins were the major tanning material for converting animal hide/skin into leather. At present, inorganic salts especially chrome salts are used to produce 70–85% leather. Replacing toxic chrome salts with less toxic and environmentally friendly tanning materials is possible by polymerization of tannins with other synthetic materials [11].

2.2.2.2 Wood composite industry

Tannins as wood adhesives have been confirmed by a number of patents. Synthetic adhesives are hazardous to environment. Therefore, tannins can be used as alternative source of adhesive in wood composite industry. Cost-effectiveness and high level of consistency should be considered for the replacement of existing harmful adhesives with tannins [11].

2.2.2.3 Water and waste water treatment plant

Tannins have a positive potentiality to be used as water treatment chemicals. For municipal water treatment, floccotan, a commercial product, has been used to eradicate suspended colloidal matter like clay and organic matter. Chelating ability of tannin is important to remove metallic ions, that is, Cr, Ni, Zn, or Cd, from waste water [11].

Acacia mearnsii cryogel obtained from the extraction of *A. mearnsii* tannin with formaldehyde in alkaline medium is used as absorbent of heavy metal (**Figure 10**). Extraction of *A. mearnsii* tannin by epichlorohydrin in N,N-dimethylformamide medium, followed by grafting with diethylenetriamine and triethylamine, yields a cationic adsorbent. This cationic absorbent is also used for heavy metals (**Figure 11**) [45].

A. mearnsii cryogel is also applicable for the removal of dyes. The dyes absorption capacity by *A. mearnsii* cryogel is presented in **Figure 12**. It is noted that cationic dyes have higher adsorption efficiency than anionic dyes, like tartrazine and alizarin violet. On the other hand, cationic adsorbent has a high efficacy for many anionic dyes, even for a cationic dye such as methylene blue (**Figure 13**) [45].

Figures 14 and **15** show the various pesticides and pharmaceutical absorption capacity of cationic tannin adsorbent. Cationic compounds, like 2,4-D, clofibric acid, MCPA, acetylsalicylic acid, diclofenac, ketoprofen, naproxen, etc., are absorbed effectively, whereas cationic products, such as atrazine, amoxicillin, trimethoprim, are adsorbed ineffectively [45].

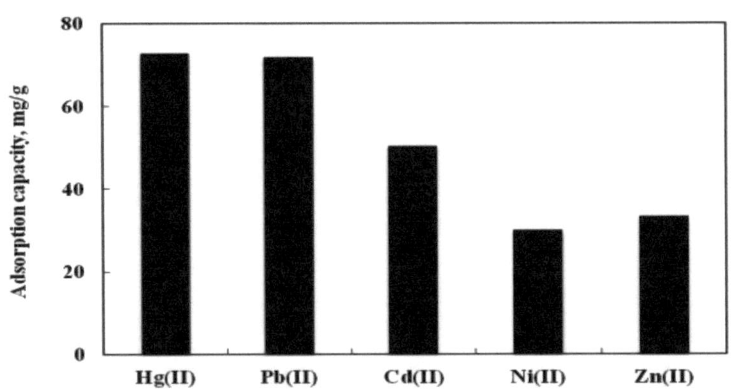

Figure 10.
Comparison of the adsorption capacity of heavy metals on Acacia mearnsii *cryogel [45].*

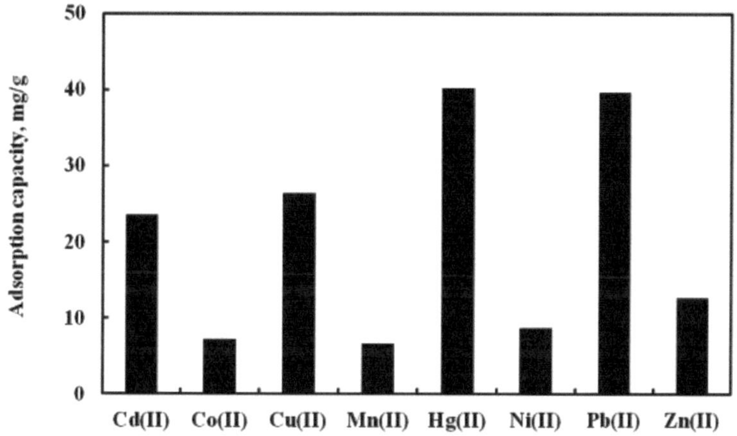

Figure 11.
Comparison of the adsorption capacity of heavy metals on cationic tannin adsorbent [45].

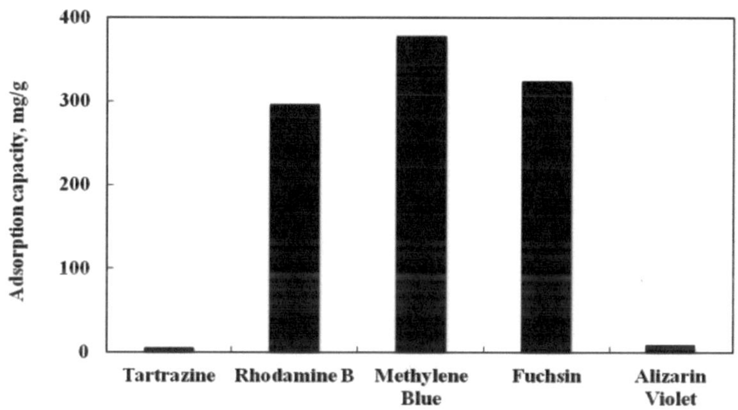

Figure 12.
Comparison of the adsorption capacity of dyes on Acacia mearnsii *cryogel [45].*

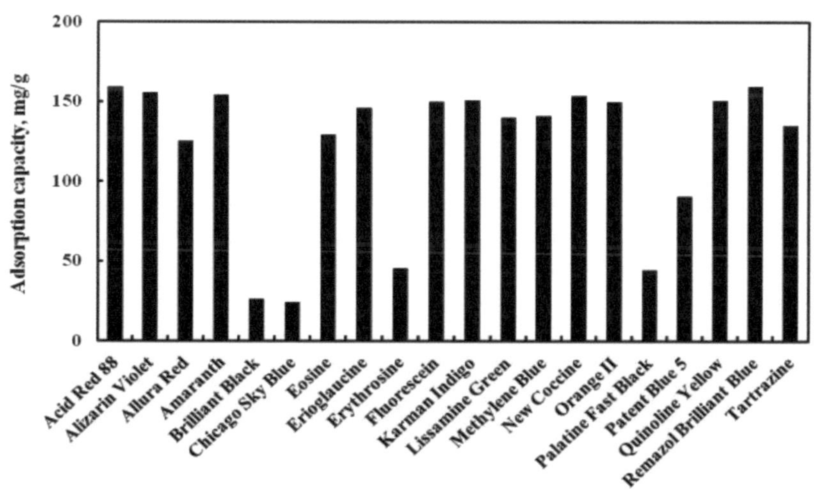

Figure 13.
Comparison of the adsorption capacity of dyes on cationic tannin adsorbent [45].

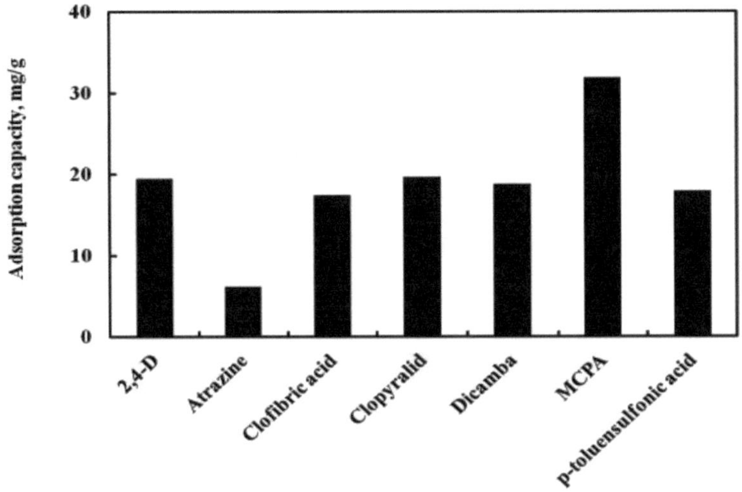

Figure 14.
Comparison of the adsorption capacity of pesticides on cationic tannin adsorbent [45].

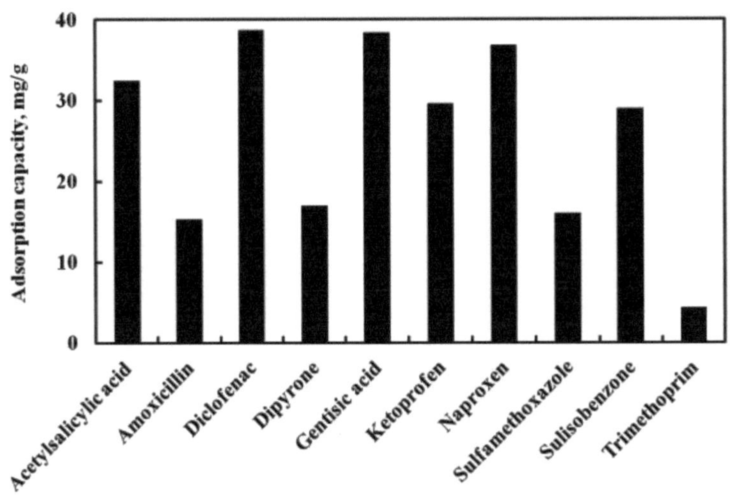

Figure 15.
Comparison of the adsorption capacity of pharmaceuticals on cationic tannin adsorbent [45].

2.2.2.4 Metal protective agents

Tannins have been used as a corrosion inhibitor for carbon steel and copper in oil and gas facilities with treated *Rhizopora mucronata* tannin. Sulfonated tannins can be used to remove scale-forming cooling water pipes and boilers. In the protective coating industry, tannins have been used for protecting metal surfaces as a primer/undercoat/topcoat. In Chile, pine tannin has been used as a commercial anticorrosive product. Black wattle extract has been used as a polyurethane-type coating for wood [11].

2.2.2.5 Building construction purpose

Tannins are used as a viscosity modifier of mud for the production of residential and architectural bricks [11].

2.2.2.6 Eco-friendly preservatives

Generally, condensed tannins preserve lignocellulosic materials naturally. Therefore, condensed tannins can be used as potential wood preservatives and biocides [46].

2.2.2.7 Medical science

Tannins have antiviral, anticancer, antibacterial, anti-inflammatory, and antioxidant properties, and these are the driving forces to be a potential source of medicinal products. A large amount of research is being undertaken currently to develop medicine from tannin [11].

Author details

Atanu Kumar Das[1], Md. Nazrul Islam[2*], Md. Omar Faruk[2], Md. Ashaduzzaman[2], Rudi Dungani[3], Enih Rosamah[4], Sri Hartati[5] and Alfi Rumidatul[3]

1 Technical Division, PT. Indiah Kiat Pulp and Paper Tbk, Serang, Banten, Indonesia

2 School of Life Sciences, Khulna University, Khulna, Bangladesh

3 School of Life Sciences and Technology, Institut Teknologi Bandung, Bandung, Indonesia

4 Faculty of Forestry, Mulawarman University, Campus Gunung Kelua, Samarinda, East Kalimantan, Indonesia

5 Faculty of Agriculture, Universitas Padjadjaran, Bandung, Indonesia

*Address all correspondence to: nazrul17@yahoo.com

References

[1] Hagerman AE. The Tannin Handbook. 2002. Available from: http://www.users.muohio.edu/hagermae/tannin.pdf

[2] Ramakrishnan K, Krishnan MRV. Tannin—Classification, analysis and applications. Ancient Science of Life. 1994;**XIII**(3-4):232-238

[3] Barbehenn RV, Constabel CP. Tannins in plant-herbivore interactions. Phytochemistry. 2011;**72**:1551-1565

[4] Khanbabaee K, van Ree T. Tannins: Classification and definition. Natural Product Reports. 2001;**18**:641-649

[5] Bianco MA, Savolainen H. Woodworkers' exposure to tannins. Journal of Applied Toxicology. 1994;**14**:293-295

[6] Bate-Smith EC, Swain T. In: Mason HS, Florkin M, editors. Comparative Biochemistry. Vol. 3A. New York, London: Academic Press; 1962. 705 pp

[7] Halam E. Practical Polyphenolics: From Structure to Molecular Recognition and Physiological Function. New York: Cambridge University Press; 1998

[8] Haslam E. Plant Polyphenols. Vegetable Tannins Revisited. New York: Cambridge: Cambridge University Press; 1989. 230 pp

[9] European patent application EP1245613A2

[10] Hillis WE. Tannin chemistry. In: Brown AG, Ko HC, editors. Black Wattle and its Utilization. Canberra: RIRDC Publication; no. 97/72; 1997. 167 pp

[11] Hobbs TJ, editor. Review of wood products, tannins and exotic species for lower rainfall regions of southern Australia. FloraSearch 1c. Report to the Joint Venture Agroforestry Program (JVAP) and the Future Farm industries CRC; RIRDC, Canberra; 2008

[12] Hillis WE. Chromatographic analysis of Eucalyptus kinos. Nature (London). 1950;**166**(4213):195

[13] Hillis WE. The chemistry of eucalypt kinos. I. Chromatographic resolution. Australian Journal of Applied Science. 1951a;**2**:385-395

[14] Hillis WE. Chromatographic examination of tannin extracts. Journal of the Society of Leather Trades' Chemists. 1951b;**35**:211-218

[15] Kirby KS, Knowles E, White T. The complex nature of vegetable tannin extracts. Journal of the Society of Leather Trades' Chemists. 1951;**35**:338

[16] Roux DG. Photometric methods of tannin analysis for black wattle tannin. Journal of the Society of Leather Trades' Chemists. 1951;**35**:322

[17] Lipsitz P, Kremen SS, Lollar RM. Untersuchung uber vegetabilische Gerbung. VII. Das Gerbungspotential gereinigter Ligninsufonate und wirklicher Gerbstoffe. The Journal of the American Leather Chemists Association. 1949;**44**:194

[18] Gustavson KH. Reaction of tetra-oxalate-diol-chromiate with hide protein. The Journal of the American Leather Chemists Association. 1947;**42**:201

[19] Cheshire A. The ageing of leather. Journal of the International Society of Leather Trades' Chemists. 1946;**30**:134-166

[20] Miekeley A. Action of acid on vegetable-tanned leather. Collegium. 1935;**786**:456-463

[21] Hathway DE. The condensed tannins. In: Hillis WE, editor. Wood Extractives. New York: Academic Press; 1962

[22] Atkins WRG. The estimation of zinc in sea water using sodium diethyldithiocarbamate. Journal of the Marine Biological Association of the United Kingdom. 1936;**20**(3):625

[23] Steven GA. Nets: How to Make, Mend and Preserve Them. Routledge; 1950. 128 p

[24] Nico R. Lab. ensayo materiales e invest, tecnol. Prov. Buenos Aires (La Plata, Rep. Arg.) Ser. II No. 38; 1950. 5

[25] Knowles E, White T. Tannin extracts as raw materials for the adhesives and resin industries. Adhesives & Resins. 1954;**10**:226-228

[26] Booth HE, Herzberg WJ, Humphreys FR. *Pinus radiata* (Don) bark tannin. The Australian Journal of Science. 1958;**21**:19

[27] Hall RB, Leonard JH, Nicholls GA. Bonding particleboards with bark extracts. Forest Products Journal. 1960;**10**:263-272

[28] Herrick FW, Bock LH. Thermosetting, exterior-plywood type adhesives from bark extracts. Forest Products Journal. 1958;**8**:269

[29] Narayanamurti D, Das NR. Tannin-Formaldehyd-Kleber. Kunststoffe. 1958;**48**:459-462

[30] Narayanamurti D, Rao PR, Ram R. Adhesives from tamarind seed Testa. Journal of Scientific and Industrial Research. 1957;**16B**(8):377-378

[31] Dalton LK. Tannin formaldehyde resins as adhesives for wood. Australian Journal of Applied Science. 1950;**I**:54-70

[32] MacLean H, Gardner JAF. Bark extract in adhesives. Pulp & Paper Magazine of Canada. 1952;**53**:111

[33] Dalton LK. Resins from sulphited tannins as adhesives for wood. Australian Journal of Applied Science. 1953;**4**:136-154

[34] Plomley KF, Gottstein JW, Hillis WE. Australia Commonwealth Sei. Ind. Research Organ. Forest Prod. Newsletter No. 234; 1957

[35] White T. In: O'Flaherty F, Roddy WT, Lollar RM, editors. The Chemistry and Technology of Leather. Vol. 2. New York: Reinhold; 1958. p. 98

[36] Anderson AB. Recovery and utilization of tree extractives. Economic Botany. 1955;**9**(2):108-140

[37] Panshin AJ, Harrar ES, Baker WJ, Proctor PB. Forest Products. New York: McGraw-Hill; 1950. pp. 469-477

[38] Farrer TW, Biek L, Wormwell F. The role of tannates and phosphates in the preservation of ancient buried iron objects. Journal of Applied Chemistry. 1953;**3**:80-84

[39] Stratta E, Ferraris GB, Genella MT. Cuoio pelli mat. Concianti. 1956;**32**:163

[40] Jurd L. Chapter 6: The hydrolyzable tannins. In: Hillis WE, editor. Wood Extractives and their Significance to the Pulp and Paper Industries. Victoria, Australia: Academic Press; 1962. pp. 229-260

[41] King HGC, White T. Tannins and polyphenols of schinopsis (*Quibracho*) species: Their genesis and interrelation. Journal of the Society of Leather Trades' Chemists. 1957;**41**:368-383

[42] Bate-Smith EC. The commoner phenolic constituents of plants and their systematic distribution. The Scientific proceedings of the Royal Dublin Society. 1956;**27**:165-176

[43] Ong KC, Khoo HE, Das NP. Tannic acid inhibits insulin-stimulated

lipogenesis in rat adipose tissue and insulin receptor function in vitro. Cellular and Molecular Life Sciences. 1995;**51**:577-584

[44] Yugarani T, Tan BKH, Das NP. The effects of tannic acid on serum lipid parameters and tissue lipid peroxides in the spontaneously hypertensive and wistar Kyoto rats. Planta Medica. 1993;**59**:28-31

[45] Combs CA. Biochemistry Research Trends—Tannins—Biochemistry, Food Sources and Nutritional Properties. New York: Nova Science Publishers, Inc.; 2016. ISBN: 978-1-63484-151-1 (eBook)

[46] Laks P. Condensed tannins as a source of novel biocides. In: Hemingway R, Karchesy J, editors. Chemistry and Significance of Condensed Tannins. Plenum Press; 1988

Chapter 2

Food Ellagitannins: Structure, Metabolomic Fate, and Biological Properties

Karen Johana Ortega Villalba, Fabrice Vaillant Barka, Carlos Vélez Pasos and Pablo Emilio Rodríguez

Abstract

Food sources of ellagitannins (ETs) are numerous, and dietary intake of these compounds is estimated up to 12 mg/day in some countries, even though ETs have been considered in the past as not bioavailable like other tannins and were mostly neglected by nutritionists. Nonetheless, new insights show that ETs are bioconverted by microbiota in the gut into metabolites called urolithins, which are bioavailable and can reach relatively high physiological concentration in the body up to 7 days after ingestion. According to the initial structure of ETs in the food source, the extent of bioconversion into urolithins may differ but all urolithins are susceptible to exert potential health benefits. Nonetheless, due to the intervention of microbiota, the production and excretion of urolithins are highly variable according to individuals, which have led to the classification of consumers into metabotype. According to metabotype, the potential health benefits of ellagitannins may differ among consumers. In in vitro, cellular and animal studies, numerous health benefits of ellagitannins and urolithins are reported mainly for the chemoprevention of hormone-dependent cancer and cardiovascular disease. Nonetheless, ellagitannins deserve closer attention from the scientific community to unravel more biological properties of this particular compound.

Keywords: ellagitannins, urolithins, microbiota, metabotype, chemoprevention

1. Introduction

Ellagitannins are food compounds that were quite neglected by nutrionists until last decade. As part of tannins, they had no good reputation and they were considered as antinutritional compounds. But new scientific insights have changed these perspectives, and ellagitannins now attract the attention of food scientists, nutritionists and consumers since the number of published papers on these compounds has considerably increased during the last decade. Ellagitannin is a hydrolyzable polymer contrary to the rest of the family of tannins and can be hydrolyzed to more simple monomers that can be eventually metabolized and that can become bioavailable with subsequent exposition of the body to these metabolites. For sure, if ellagitannins are widely present in nature, only few food sources are reported with relatively high content of this compound, and consequently exposition of consumers to food ellagitannins is relatively low, especially in the Western diet. But given the health potential of ETs, ET-rich food now belongs to the select group of functional foods, and their consumption should be considerably enhanced in the future. Actually, given the main food source of ellagitannins such as berries and nuts, we can easily assume that the exposure to this compound and their metabolites was considerably higher in the hunter-gatherer diet than in modern time. Without presuming that an increase in ET intake would reduce significantly the impact of certain chronic diseases due to modern lifestyle, it is reasonable to argue that ETs have been part of our evolutionary history and they could potentially perform health care functions. New scientific insights presented in this chapter on the in vivo metabolisms of ellagitannins and the potential biological activities of generated metabolites tend to support this hypothesis. This review presents the main food source of ellagitannins, their general chemical structure, and how technologies and storage could eventually affect ellagitannin composition in processed foods. Then, we will review the metabolomic fate and the bioavailability of ellagitannins in humans, which is strongly related to the performance of intestinal microbiota, and finally, we will present a summary of the main biological activities, attributed to ETs and their derived metabolites.

2. Food occurrence of ellagitannins

Ellagitannins (ETs) are with gallotannins part of the hydrolyzable tannins and constitute the largest group among more than 500 hydrolyzable tannins characterized until now [1]. To date, more than 1000 natural ellagitannins have been identified in nature [2] but most of them are not preponderant in foods. The main ETs identified in foods (specially in fruits, nuts, and seeds) are punicalagin, sanguiin H6, lambertianin C, pedunculagin, vescalagin, castalagin, casuarictin and potentillin (seeds) [3].

Examples of concentration and ellagitannins identified in some foods are presented in **Table 1**. The occurrence of ETs in foods is restricted to a few fruits, such as berries of the genus *Rubus* (cloudberry, raspberry, blackberry, blueberry, and cranberry) and the genus *Fragaria* (strawberry), pomegranate, nuts (walnuts and almonds), seeds, and oak-aged wines [1, 6–8]. Recently, other ET food sources of

Food source	Ellagitannins	Content equ. EA (mg/100 g)
Blackberries (*Rubus* spp..)	Sanguiin H6, lambertianin D [7]	150–270 [3]
Strawberries (*Fragaria ananassa*)	Casuarictin, pedunculagin, sanguiin H6 [7, 8]	71–83 [3]
Cloudberries (*Rubus chamaemorus*)	Sanguiin H6, lambertianin C [7]	312 [3]
Raspberries (*Rubus idaeus*)	Sanguiin H6, sanguiin H10, lambertianin C [7, 9]	326 [10]
Pomegranate (*Punica granatum*)	Punicalagin [7]	58–177 [3]
Guava (*Psidium* friedrichsthalianum)	Pedunculagin, castalin, and vescalin [5]	63 [5]
Jabuticaba (*Myrciaria jaboticaba*)	Sanguiin H6-H10, lambertianin C [4]	900 [11]
Muscadine grapes (*Vitis rotundifolia*)	Sanguiin H5 [7]	3–91 [7]
Purple Grumixama cherry (*Eugenia brasiliensis*)	Pedunculagin, strictinin, castalagin, vescalagin [12]	16 [12]
Chestnuts (*Castanea sativa*)	Castalagin [7]	149 [13]
Pecans (*Carya illinoinensis*)	Pedunculagin [7]	316 [13]
Walnuts (*Juglans regia*)	Pedunculagin, casuarictin [14]	864 [13]

Table 1.
Main food source of ellagitannins.

local importance have been identified such as jabuticaba [4], guava, [5] and grumixama cherries [12]. It is interesting to note that according to a Brazilian research team, Jabuticaba berries from a particular variety cultivated in south Brazil have the highest registered ET content in fruits. Berries have almost three times more equivalent EA content than walnuts and pecans and at least 15 times more than other fruits and nuts [6]. In the berries from the genus *Rubus* and genus *Fragaria*, total equ. EA content represents the most important compounds with 50–88% of total phenolic. Also, ET content can be considerably affected by variety, ripeness, fruit parts, geographic origin, climate, season, cultural practices, and mineral nutrition [6].

ET daily intake is generally low and has been estimated around 5 mg/day for Western diets with major contributors being the red berries mainly strawberries, followed by raspberries and blackberries. Given the significant seasonality of the production of these fruits, the exposure to ellagitannins is very uneven during the year. In the Scandinavian countries, where the consumption of berries increases considerably in summer, daily intake can reach up to 12 mg/day [6, 9] with cloudberry, raspberry, rose hip, strawberry, and sea buckthorn being the main contributors with content from 1 to 330 mg/100 g (fresh weight basis).

3. Chemical structure of ellagitannins

ETs are esters of hexahydroxydiphenic acid (HHDP) and a polyol, usually glucose or quinic acid [3, 15, 16], that when they are hydrolyzed spontaneously suffer lactonization to form ellagic acid [10, 17] (**Figure 1**).

Chemical structure diversity among ETs is huge and is due to the possible variations in position, frequency, and stereochemistry of the HHDP units, galloylation extent, and/or anomeric stereochemistry of sugar moieties [17]. Thus, due to seemingly endless structural variations among ellagitannin, elucidating their native structure is often a challenge [17]. According to their chemical structure, ETs can readily undergo different chemical reactions such as transformation, isomerization and oligomerization, which finally determine overall physico-chemical properties, hydrolytic susceptibility, and finally biological activity in vivo [7].

The important structural diversity of ET structure is due to the different possible extent of galloylation and formation of aromatic C-glycosides, the number of intramolecular C-C coupling of galloyl groups and hydrolytic cleavage of galloyl-derived aromatic rings, the level of dehydrogenation, and oligomerization via oxidative C—O [2, 17]. According to the number of HHDP groups linked to sugar moiety, ETs can be classified into monomeric, oligomeric, and polymeric ellagitannins [18].

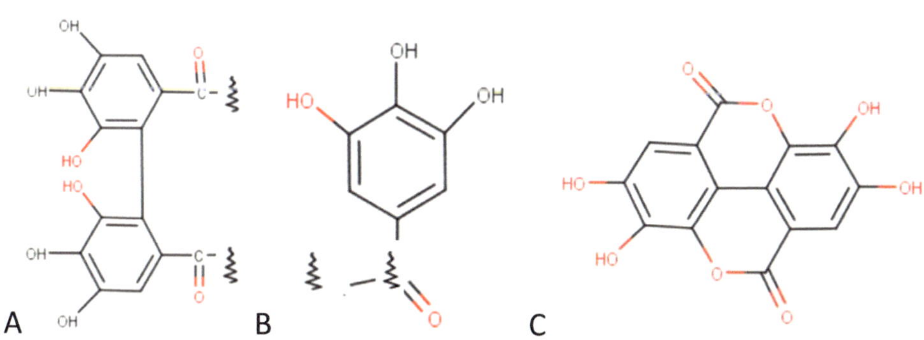

Figure 1.
Basic structures of ellagitannins: (A) HHDP acid (R radical); (B) galloyl unit (G radical); (C) ellagic acid.

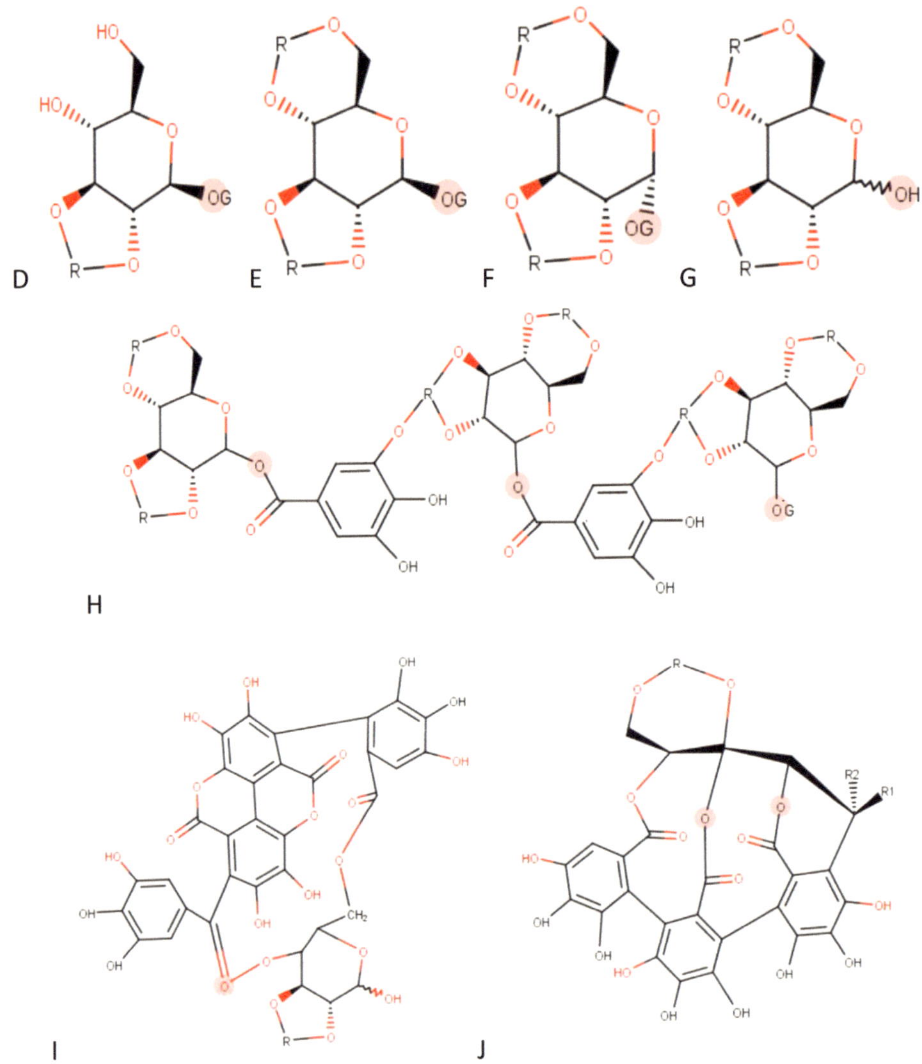

Figure 2.
Most common ellagitannins present in food: (D) sanguiin H5; (E) casuarictin; (F) potentillin; (G) pedunculagin; (H) lambertianin C; (I) punicalagin; (J) vescalagin R1: OH, R2:H o Castalagin R1:H, R2:OH.

ETs are generally hydrolyzable in acidic or basic solutions. Even though ETs are quite resistant to acid hydrolysis, neutral or slightly alkaline pH (from 7.0 to 7.3) are the best conditions for ET hydrolysis to occur [19]. During hydrolysis, the ester bounds in the polymer are cleaved and HHDP is released followed by a spontaneous lactonization of free HHDP unit into free ellagic acid or derivatives [20]. This reaction is mainly used for the detection and quantification of ellagitannins, as ET content in food samples is often expressed as ellagic acid equivalents (**Figure 2**). Acidic and basic hydrolysis of ETs can also occur during food processing, storage, and passage through the stomach and duodenum [10, 16, 21]. Although most ETs are hydrolyzable, further C—C coupling of polyphenolic residue with the polyol unit, such as in the case of vescalagin, can prevent hydrolysis [22]. In addition, ETs can undergo polymerization reactions during maturation of fruits or thermo-physical treatments, which can make them insoluble and eventually attached covalently to cell wall fragments [21]. Nonetheless, most often acidic or basic hydrolysis will allow generating ellagic acid (EA) from ET even in mild conditions. As a consequence, EA, the dimeric derivative of gallic acid, is often spontaneously present in its free form in plants, next to ET [23].

3.1 Structural changes of ETs and EA during process and storage

Different studies have pointed out that there are important changes in ETs' composition during processing and storage with marked subsequent consequences on the bioavailability and bioactivity of these compounds [24, 25].

Mazur et al. [26] noticed a decrease of 7% of lambertianin C in red raspberry jam after 6 months of storage at 20°C in dark; meanwhile, ellagic acid derivatives and total phenolics increased by 47%. The reason was the spontaneous hydrolysis of ellagitannins to ellagic acid that may occur during storage. Authors showed also that according to the genotype of raspberry, losses of ET were more or less important. On the other hand, one ellagitannins, the sanguiin H6, remains remarkably stable during storage in this case [26]. Nonetheless, in another study, stored pasteurized blackberry juice showed after 6 months at 25°C a 46% loss in sanguiin H6/lambertianin A, 42% loss in lambertianin C, and 72% loss in lambertianin D. Like previously, the total amount of EA measured after hydrolysis registered only minimal changes, which evidenced the spontaneous depolymerization of ETs into EA during storage. At 5°C, half-life ($t_{1/2}$) of sanguiin H6 and lambertianin C was almost comparable (around 80 days), but when stored at 45°C, $t_{1/2}$ was four times higher for sanguiin than for lambertianin [25]. Sanguiin H6 is a dimer and lambertianin C a trimer, which could explain the higher stability of this compound during storage at relatively high temperature. At freezing temperature, no changes at all were observed over a 6-month period [24]. We can conclude that stability of ellagitannins during storage of processed foods depends on their chemical structure as well as the composition of the food matrix.

Between different processing alternatives evaluated by Hager et al. [24], canning, pureeing, and freezing had little effect on blackberry (cv. Apache) ellagitannins, but the removal of seeds in the press-cake generates a loss of up to 70%, and if juice is microfiltered, 12% more loss can be registered. In that case, some ellagitannins appear to be associated with cell fragments and with the mucilage that surrounds the seeds. In the case of Costa Rican guava, pressing appears to increase the content of some ETs. After pressing, an increase in pedunculagin isomer 1 of 25% was observed, while castalagin isomers presented a significant decrease (40%). Authors suggest that castalagin decrease may be linked to degradation and not to isomerization of the compounds, since no increase in vescalagin content was observed [5]. Milling appeared to enhance the content of pedunculagin isomer 1 and castalagin isomers by 34 and 31%, respectively. Actually, another study on the effects of mechanical and enzymatic pretreatments on the extraction of ellagitannins from blackberries showed that enzymatic treatment (pectinase and cellulase) combined with continuous pressing enhanced significantly ET content in the juice (from 437 to 982 mg ellagic acid equivalents/100 g (dry basis)) [27].

Critical steps on ET content during classical industrial processing of blackberry-based beverage in glass bottles were evaluated. Hot-filling that is characterized by long-term exposure of the beverage to high temperature was the operation that most degraded ellagitannins with losses of 80% in lambertianin C and 50% in sanguiin H6 in the final product. Again, sanguiin H6 showed a higher thermal stability than lambertianin. It was also observed that the intensity of thermal treatment during process affects stability of ETs during storage [25]. But, when dealing with equivalent total EA content after hydrolysis, no changes were observed. On the other hand, in another food source, the ripe Costa Rican guava (*Psidium friedrichsthalianum* Nied), different pasteurization treatments (71.1°C for 4 s and 60°C for 8.2 min) were found to not affect ETs' final content. In this case, the composition in geraniin, vescalagin, and pedunculagin isomers remained basically constant during the process [5]. The analysis of these results shows that thermal stability at high temperature is also affected by the chemical structure of ET.

In other processing operations such as osmotic dehydration in 50–65°Bx sucrose solutions (30°C), in the case of blackberry, 80% of ellagitannins were retained after 1 h, while losses reached up to 45% after 3 h. The concentrations of the two main ellagitannins, lambertianin C and sanguiin H6, revealed similar patterns of variation. On the contrary, due to a much lower molecular weight, the loss of free ellagic acid reached up to 50% after 1 h of osmotic dehydration [23].

4. Metabolomic fate and bioavailability of ETs and EA

Many studies have proven that ETs are not bioavailable as such, and they have never been detected in human plasma after normal consumption of ET-rich foods [28, 29]. Ellagitannins are probably not bioavailable because of their size (above 634 Da for one of the simplest ellagitannins, the sanguiin H4 ($C_{27}H_{22}O_{18}$), up to 3740 Da for lambertianin D ($C_{164}H_{106}O_{104}$)). Their relatively high polarity and the presence of a C—C linkage could also explain this situation. During ingestion, ETs can also bind some proteins in saliva and cause astringency, and in this case, they may not be metabolized further [30]. Also, some ellagitannins are resistant to acid and basic hydrolysis in the GI tract and they can reach almost intact the large intestine [31]. However, for most ETs sensitive to acidic and basic hydrolysis in the stomach and the duodenum, respectively, they release free ellagic acid, which is at its turn poorly bioavailable. In the human digestive system, the bioavailability of EA derivatives depends on the part of gastrointestinal tract in which these compounds are released [7]. In stomach or small intestine, only low level of absorption could occur, and if EA can be detected in plasma and urine between 1 and 5 h after ingestion of dietary ETs, generally as methyl and dimethyl ethers or glucuronic acid conjugates, all these EA derivatives are always found at very low concentrations [29, 32, 33]. Low bioavailability of EA is probably due to low water solubility, and to its ability to bind irreversibly to cellular DNA and proteins, or to form poorly soluble complexes with calcium and magnesium ions which affects transcellular absorption [3, 33, 34].

The extent of the degradation of ETs in the upper GI tract depends on their chemical structure, the food matrix, and their susceptibility to acid/base hydrolysis in the stomach and duodenum [29]. Therefore, some ETs can reach the intestine where they can exert potential biological activity and eventually some can be partially converted into EA by enzymes from microbiota [35].

In the lower GI tract, released EAs can be partially metabolized by gut microbiota into urolithins (Uro, dibenzopyran-6-one metabolites) through reduction of one of the two lactone groups followed by decarboxylation and sequential dehydroxylation involving a step-by-step reduction to tetrahydroxy (urolithin D), trihydroxy (urolithin C), dihydroxy (urolithin A and isourolithin A), and monohydroxy dibenzopyranones (urolithin B) (**Figure 3**) [21, 35–37].

Urolithins appear to be the main plasma and urinary biomarker after consumption of ET-rich food (**Table 2**). Specifically, urolithin A and B and their phase II metabolites are the main metabolites in plasma and urine [10, 13] detected at micromolar concentration level. They can also be found at much higher concentration in some tissues or organs such as prostate gland and colon where they can accumulate [38, 39]. Persistence of urolithins in the body has been reported for long periods

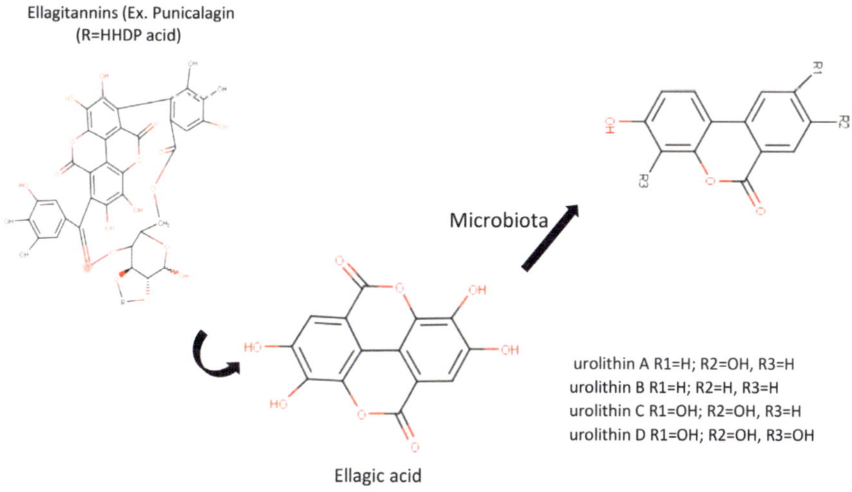

Figure 3.
Metabolism of ellagitannins in the GI tract.

after a single intake of ET-rich food. Urolithins in urine have been detected up to 7 days [58] after dietary ET intake, and this long persistence in the body is attributed to the involvement of microbiota and enterohepatic recirculation [9, 40–42].

Therefore, urolithins have been proposed to be responsible for biological activity of ETs [38] as they remain in the body at relatively low concentrations but during a long time with potential homeopathic-like effect. However, the concentration of urolithins in plasma, urine, and feces varies considerably between individuals [38, 46, 47]. Actually, the huge inter-individual variability of microbiota composition affects the production of urolithins, which is mediated by microbiota. During different studies, some individuals were labeled "low excreters" of urolithins after ingestion of ET-rich food, while other individuals were labeled as "high excreters of urolithins A or B." Therefore, recently, a stratification of individuals according to their urolithin excretion status in urine has been proposed. Three metabotypes were defined: metabotype A, which includes main excreters of urolithin A; metabotype B with main excreters of urolithin A and B; and metabotype 0 corresponding to low urolithin excreters [31]. This classification appeared to be consistent across multiple intervention studies, independent of the ET food source, and health status of participants [46]. The distribution of urolithin metabotypes in adult population varies probably according to geography, but in a Western adult population taking into consideration a large cohort of individuals, UM-A is the most abundant metabotype with 55% followed by UM-B (34%) and UM-0 (11%) [47]. Nonetheless, a recent study reported by Cortés-Marín and coworkers [47] showed for the first time in a Caucasian cohort (5–90 years, n = 839) that age could determine the individual's capacity to metabolize EA into urolithins A and B, even though the percentage of the population with low ability to excrete urolithins (metabotype 0) remains around 10% of the population whatever the age considered. The percentage of individual with metabotype A was higher in the case of children (80%) and decreased steadily after adolescence while metabotype B increased [47]. Authors also reported a significant association between increased physical activity and prevalence of UM-B especially between 5 and 18 years. On the other hand, no correlation of a specific metabotype with gender, body mass index, weight, health status, and diet was observed [47].

Nonetheless, most studies tend to show a strong persistence of metabotype status for adults even though recent research showed that individuals with UM-0 status have managed to become urolithin excreters of UM-A or UM-B after a long-term

Source ETs	Study design	Metabolites	Main observations
Red raspberry (300 g) single dose [9]	9 adults: 5 females and 4 males (22–44 years old) sampling: blood (0, 0.5, 1, 1.5, 2, 3, 4, 5, 6, 8, 24 h), urine (all until 48 h).	In urine, UA aglycone and phase II metabolites, In plasma: UA and UB and phase II derivatives dimethylellagic acid-O-gluc.	Urolithins A and B, present C_{max} in plasma in times >24 h, and in urine between 32 and 48 h after intake, showing high persistence in the body.
Pomegranate extract (1.8 g in 4 capsules) single dose [43]	10 men and 10 women (20–22 years old) sampling: blood (0, 0.5, 1, 2, 3, 4, 5, 6 and 24 h), urine 0, 24, 48 and 72 h.	In urine and plasma, UM-0: 2 people; UM-A: 16 people and UA were quantified after 24, 48, and 72 h); UM-B: 2 people (UA and UB).	EA bioavailability was affected by the presence of ellagitannin, pH, and protein as well as low pH. A higher free EA intake does not enhance bioavailability of EA but promotes urolithin production.
Strawberry extracts rich in: (1) free EA; (2) monomeric ETs, and (3) dimeric ETs (0.1% single intake) [37]	30 Wistar rats (10 each treatment), sampling: feces, urine and blood, intestinal digesta (2, 4, 7 days after intake); samples of digesta, stomach, small intestine, cecum.	In urine, plasma, and intestinal digesta: UA, UB, UA gluc, Nasutin A gluc, EA dimethyl-ether-gluc (DMEAG). Feces: Nasutin A and UA.	ETs and EA metabolites are found in gastrointestinal digesta, blood plasma, and urine, 2 days after intake. UA and DMEAG were founded in rats fed with free EA extract, Nasutin A and UA and Nasutin
Black raspberries (freeze-dried, 10% w/w, 6 weeks) [44]	12 male mice, sampling: blood and luminal, colon, liver, and prostate tissue.	Urolithin A (all mouse plasma, liver, prostate, colon, and luminal tissue), urolithin C (all tissues except prostate), urolithin B and urolithin D (only in luminal tissue).	Highest amount of UA was found in luminal tissue. UC was 45-fold more abundant in colon with respect to plasma. UB and UD were found in very low amounts with respect to UA. Luminal contents presented lower abundance of *Clostridium* with respect to *Barnesiella* in mice after intervention.
Pomegranate extract (PE) (1.8 g/day, 3 weeks) [45]	Simulator of the human intestinal microbial ecosystem TWIN-SHIME®, coupled to human colon adenocarcinoma cell line Caco-2 cells (HTB37), UM-A and UM-B were incubated.	Large intestine: IsoUA and UB, detected only in UM-B, UA in UM-A and UM-B.	Production of urolithin starts earlier and faster in UM-B than in UM-A. Chronic PE intake improves UA production in both metabotypes, showing similar profiles after 18 days. UA production was directly related with Gordonibacter abundance.

Table 2.
Some relevant studies on the metabolism of Ellagitannins-rich foods.

exposure (up to 6 months) to a high ET source, a pomegranate-concentrated extract [48]. Thus, it seems that microbiota ability to catalyze the production of urolithins could be influenced by long-term exposure to ET-rich food by promoting growth of the bacteria involved in the urolithin metabolism [40]. Nonetheless, it was not reported if the change of metabotype remains persistent after the end of the study, when diet comes back to normal.

Recently, an attempt to find a correlation between urolithin metabotypes and enterotypes of the human gut microbiome proposed by the Human Microbiome Project (HMP) [39] has been made by Romo-Vaqueroa and coworkers [49] following a cohort of 249 healthy volunteers after walnut or pomegranate extract ingestion for 3 days. Results showed that urolithin metabotypes and enterotypes (enterotype 1 (preponderance of Bacteroides), enterotype 2 (preponderance of Prevotella), or enterotype 3 (preponderance of Ruminococcus)) were not coincident. Only for enterotype 2, UM-A was slightly higher than UM-B. Nonetheless, a higher diversity of microorganisms in UM-B individual with respect to UM-A, and even more with respect to UM-0, was found. Actually, urolithin B production requires a more complex enzymatic arsenal than urolithin A. It was observed that a higher relative importance of microorganisms from the Coriobacteriaceae family tends to be correlated with a higher preponderance of UM-B with respect to UM-A and even more with respect to UM-0. Two bacterial strains isolated from human microbiota, *Gordonibacter urolithinfaciens* and *Gordonibacter pamelaeae* from Eggerthellaceae family, which was previously considered as part of Coriobacteriaceae family, showed ability in vitro to transform EA into urolithin C [50]. Recently, a specific strain also isolated from human microbiota was able to further metabolize EA up to isourolithin-A and was named as a consequence *Ellagibacter isourolithinifaciens* [50, 51]. Nonetheless, until now, no other bacterial strains that could metabolize EA to urolithins B have been reported.

Another factor that can impact the rate of urolithin production in vivo is the food source and the chemical structure of ingested ETs. For example, more urolithins in prostate from patients who consumed walnuts rather than in patients who consumed pomegranate juice, even when the latter had a higher ET content, were found. [14]. Also, it appears that there is a kind of saturation of the metabolic pathways as the amount of urolithins excreted in vivo remains apparently independent of the quantity of ETs ingested. This was observed for different food sources of ETs consumed in normal quantities such as strawberries, raspberries, walnuts, and oak-aged red wine [52]. Probably, there exists a consumption threshold below which urolithin excretion cannot be detected.

At last, the bioavailability of ETs and EA could be affected by food processing. In a clinical study, 16 healthy volunteers consumed approximately the same quantity of ET but in different presentation: pomegranate juice (PJ), pomegranate polyphenol liquid extract (POMxl), and pomegranate polyphenol powder extract (POMxp). As a result, there were no statistical differences in the level of EA in plasma between the three interventions over a 6-h period. Only, POMxp presented a longer lag-time to reach the peak of maximum concentration compared to PJ and POMxl [53]. A similar study was performed with 20 healthy volunteers comparing pasteurized strawberry juice (80°C for 5 min) and the equivalent fresh fruits. In this case, processing did not affect the urinary excretion of urolithins. Although the amount of free EA was increased 2.5-fold during processing, no effect on the urinary excretion of urolithins was observed [40]. Actually, further researches are required for a more definitive assessment on the effect of processing on the production and excretion of urolithins.

5. Main biological activities of ETs and their metabolites

Ellagitannins, ellagic acid, and their metabolites have been reported to exhibit numerous beneficial effects on human health including anti-inflammatory, anticancer, antioxidant, prebiotic, and cardioprotective properties [21, 54]. However, in vitro studies with cells or in vivo studies with animals could give inconsistent or untranslatable information about bioactivity of these metabolites in humans. Abundant literature shows for example the impact of ETs and EA on cells from organs that are not part of the GI tract and the results are absolutely controversial and inconsistent with actual knowledge. The potential health effect of ETs and possibly EA can only be exerted within the GI tract, as these compounds are poorly bioavailable. On the other hand, urolithin production is mediated by microbiota and studies on animals can hardly be extrapolated to humans, except in the case of germ-free animals used in human microbiota-related researches. Therefore, in this review, even though research studies are much more scarce, we will report results of biological activities of ETs and EA only related with the GI tract, and for urolithins, given the importance of microbiota in their metabolism, we will focus only on the results of clinical trials with humans.

5.1 Effect of ETs and EA in GI tract

In in vitro model of colon cancer, ellagic acid was found to have a significant antiproliferative effect inducing apoptosis of cancer Caco-2 cells via a mitochondrial pathway and without side effects on normal colon cells [55]. In another study in rats, EA showed anti-inflammatory properties by iNOS, COX-2, TNF-α, and IL-6 downregulation due to NF-κB repression. Authors conclude that EA may exert a chemopreventive effect on colon carcinogenesis [56]. Furthermore, ETs and EA have a high antioxidant activity (even higher than urolithins) and could be highly efficient to scavenge oxygen free radicals and eventually prevent inflammation and colon cancer [21]. At last, in nasopharyngeal carcinoma cell lines (NPC-BM1), EA has showed ability to downregulate Bcl-2 and DNA fragmentation, by increasing caspase-3 enzymatic activity, which reduces telomerase activity [57].

5.2 Effect of urolithins

Numerous studies have demonstrated the metabolism of EA into urolithins, in approximately 12–24 h, with persistence of urolithins up to 4–7 days in urine after dietary intervention [9, 41, 42, 58], and research interest has shifted to the potential effect of urolithins on health. Actually, urolithins may exert a much more consistent effect at the systemic level than EA with concentration in body fluids at least one order of magnitude higher in body fluids. Actually, EA has been reported in plasma at concentration around some nanomoles per liter, while urolithins have been reported at concentration level up to 80 μmol/l [58].

5.2.1 Anti-inflammatory effect

Inflammation is a primary defensive response against harmful factors that involved different mechanisms including the immune system cells [8]. The impact of urolithins on inflammatory processes has been well established on various in vivo and in vitro models [59–61]. In in vitro models, urolithin aglycones have been tested, while in vivo glucuronide conjugates of urolithins are the predominant metabolites present in plasma, tissues, and urine [62]. Urolithin aglycone is highly bioactive against inflammation, but some authors suggest that urolithin conjugate may be even more active. In a study, in which urolithin conjugates (iso-Uro-A-gluc, Uro-A-gluc, and Uro-B-gluc) were isolated from urine of a volunteer after ingestion of pomegranate juice (0.5 L/day), walnuts (30 g/day), hazelnuts (30 g/day), and fresh raspberries (200 g/day) for 5 days, the cleavage of glucuronides by endogenous β-glucuronidases released by human neutrophils was observed . β-Glucuronidase is an enzyme that is released from inflammatory cells and the lysosomes of necrotic cells, and high levels can be found in most solid tumors. Therefore, a wide number of structurally diverse glucuronide prodrugs have been designed with the aim of enhancing the selectivity of cancer chemotherapy. The results suggest that the selective activation of urolithin glucuronides by β-glucuronidase could locally increase the concentration of bioactive urolithin aglycones. More clinical trials are needed to better understand the anti-inflammatory response attributed to urolithin [62] and the impact on the suppression of the immune responses, especially on inflammation-associated diseases, like cardiovascular diseases and cancer [63].

5.2.2 Chemoprevention of cancers

Several in vitro and in vivo (animal or human) studies have reported a protective effect of urolithins on prostate cancer [64–66]. However, recent clinical interventions have pointed out inconsistent results. A recent review compiles the data of clinical trial studies after consumption of pomegranate juice or extracts and discusses whether urolithin could inhibit or slow the growth of prostate cancer in patients. The authors reported a significant increase in urolithin A in prostate but a nonsignificant reduction in 8-hydroxy-2-deoxyguanosine, a marker of oxidation in cancer tissue, for neoadjuvant patients subjected to radical prostatectomy after pomegranate intake (1200 mg polyphenols/day) compared with placebo in a large trial (4 weeks, n = 33) [67]. Similar results with muscadine grape skin extract have evidenced no benefit on recurrent prostate cancer patients in spite of urolithin A increase [68]. Nonetheless, authors [67] have noted in a specific group of patients (named AA genotype), which has been previously associated with more hostile prostate cancer and more sensitivity to antioxidants, a significant increase in a prostate-specific antigen doubling times (PSADT), an antigen that is claimed to slow tumor growth.

After an acute intake of grumixama cherry juice (Brazilian) by healthy women (n = 10), the antiproliferative activity of urolithins against breast cancer cells (MDA-MB-231) was evaluated. The extracts of urine exhibited seven urolithins, mainly urolithin C and urolithin A. Those extracts obtained during 2–4 h after juice intake presented the highest inhibition of proliferation of MDA-MB 231 breast cancer cells. This inhibition was attributed to a significant G2/M cell cycle arrest (apoptosis) occurred in MDA-MB-231 cells, and was demonstrated by the increase in sub-G0/G1 populations. Additionally, the authors linked this inhibition to a possible synergy among anthocyanins and urolithins [12]. Then, the modulation of the positive-estrogen receptor in breast cancer cells (MCF7) is probably one of the potential actions of urolithin conjugates. On the other hand, in a study realized with breast cancer patients (n = 19) who consumed a mixed extract (493.4 mg phenolics/day) containing pomegranate, orange, lemon, olive, cocoa, grape seed extracts plus resveratrol, theobromine, and caffeine, it was shown that the main metabolites detected in breast tissues were urolithin-A-3-O-glucuronide. Nonetheless, no antiproliferative or estrogenic/antiestrogenic activities in MCF-7 breast cancer cells were reported [69].

Ellagitannin gut microbiota-derived metabolites have shown a wide range of colon anticancer effects both in cellular and animal studies [70–72]. However, the current clinical evidence that confirms their colorectal cancer (CRC) chemopreventive effect in humans is still very weak. A study evaluated the modification of microRNAs (miRs) expression, one CRC biomarker, in normal and malignant colonic tissues from CRC patients after pomegranate extract intake (900 mg/day before surgery). As a result, pomegranate consumption seems to moderate the modulation of various specific miRs in colon tissue, but there was no association between tissue urolithins and the detected miRs changes, which were attributed to a possibly critical surgery alteration in miRs levels that did not allow to discriminate between malignant and normal tissues [73]. Another more recent study in 35 patients with colorectal cancer (CRC), daily supplemented with pomegranate extracts, was conducted to evaluate the expression of various CRC-related genes in normal and cancerous colon tissues. Before (biopsies) and posterior (surgical samples) to pomegranate intake (5-35 days). Despite the consumption of pomegranate extract was significantly associated with a balancing effect in the expression of genes regulated by the experimental protocol, these results were not associated with the individual metabotypes or the levels of urolithins and EA in the colon tissues. Consequently, the in vitro effects were not reproduced in vivo evidencing discrepancy between results [70]. In general, we can conclude that there is a lack of clinical interventions with ET-rich food in humans; besides, these kinds of studies are essential to corroborate the real effect of urolithins on cancer.

5.2.3 Reducing risks of cardiovascular disease (CVD)

The gut microbiota is frequently presented as a key factor in the evolution of obesity and cardiovascular disorders (CVD) [74]. One clinical study clustered urolithin metabotypes (UMs) of 18 healthy overweight/obese subjects with the aim of correlating metabotype status with CVD biomarkers after pomegranate extract consumption. In baseline and before UM clustering, the whole group exhibited mild dyslipidemia, and after clustering, only the serum lipid profile of UM-B individuals (n = 15) showed moderate risk values in total cholesterol, intermediate-LDL-cholesterol, as well as other serum lipids related to CVD risk. After ET intervention, only blood biomarkers of UM-B subjects were improved after pomegranate extract intake, reducing their CVD risk. Interestingly, a dose-dependent behavior was notable only in UM-B patients [48].

Another experiment comparing healthy patients with patients with metabolic syndrome (MetS), both consuming walnuts, showed that urolithin A only was inversely correlated with glycaemia in MetS individuals. Additionally, when MetS patients with UM-A were treated with statin, their lipid profile became similar to healthy individuals. This was not the case for individuals with UM-B [74].

Another study showed that the increasing relative importance within the microbiota of bacteria from the Coriobacteriaceae family such as *Olsenella, Senegalimassilia, and Slackia*, which characterized UM-B status, was positively correlated with blood cholesterol levels and normal BMI [49].

Endothelial dysfunction and inflammation are both usual events that occur in the development of atherosclerosis. The correlation between the plasma urolithin metabolites and improvement in endothelial function after red raspberry intake was reported. Endothelial function measured as flow media vasodilation (FMD) presented two peaks, first at 1–2 h after intake, linked with EA plasma peak concentration, and second peak at 24 h, associated with urolithin-3-glucuronide and urolithin-A-sulfate absorption peaks. Similar results were reported by other authors in cranberry and blueberry juice interventions [75], but it was shown that effect was the same when consuming 200 or 400 g of raspberries. Additional distinctive key factors in atherosclerosis development have been reported, as the capacity of monocytes to adhere to endothelial cells and the uptake and efflux of cholesterol by macrophages. In vitro, urolithins and EA were able to reduce adhesion of THP-1 monocytes to human umbilical vein endothelial cells and reduce secretion of sVCAM-1 and IL-6, a cellular adhesion molecule and a pro-inflammation cytokine, respectively. Also, urolithin C and EA were associated with decreased accumulation of cholesterol in THP-1–derived macrophages [76]. Attenuation of THP-1 also was reported in the presence of urolithin A in endothelial cells and also reduced considerably the expressions of ICAM-1 and MCP-1, an intercellular adhesion molecule and a monocyte chemotactic protein, respectively [77].

6. Concluding remarks

Ellagitannins are present in considerable amounts only in some specific food sources such as berries and nuts, but some tropical fruits deserve attention. Their diverse structure can be modified during food processing resulting in free ET and EA derivatives, which are poorly bioavailable. After ingestion, most ETs are spontaneously converted into EA, which is poorly bioavailable and can be used as substrate by gut microbiota. The main products resulting from the action of gut microbiota on EA are urolithins. The main biomarkers in blood and urine of ET-rich food exposure are urolithins A and B. Nonetheless, there is an important interindividual variability in the excretion of urolithins, and this observation has led to the classification of the population in three metabotypes: the "low urolithin" excreters that represent approximately 10% of the population; the "urolithin A" excreters, the most important group with approximately 55% of individuals; and finally, the "urolithin A and B excreters" that represent around 35% of the population. The metabotype status appears to be quite persistent although it can change during life span and the constant exposure to ET-rich food appears to increase urolithin production. Microorganisms from the Coriobacteriaceae family were identified as urolithin producers and the relative importance of this family within the microbiota was apparently correlated with the metabotype. The stratification of individuals by their metabotype was essential to overcome inconsistencies during clinical trial, and it must be taken into account in all future intervention studies. The positive biological effects of ETs and EA at the level of the GI tract are consistent and reported by various authors. For urolithins, the panorama is more confused, and more long-term clinical intervention studies with human are required. Nonetheless, at the end of this review, the potential health effect of ET-rich foods is definitely promising and they deserve to be part of a healthy diet as functional foods.

Acknowledgements

This publication was jointly financed by the International Research Center in Agronomy for Development (CIRAD), Montpellier, France, and the Colombian Research Cooperation in Agronomy (AGROSAVIA), Colombia. Karen Johana Ortega is a PhD student with a scholarship granted by the Colombian government through Colciencias 727-2015 announcement.

Conflict of interest

The authors confirm that there is no conflict of interest.

Author details

Karen Johana Ortega Villalba[1], Fabrice Vaillant Barka[2,3]*, Carlos Vélez Pasos[1] and Pablo Emilio Rodríguez[2]

1 Universidad del Valle, Cali, Colombia

2 Colombian Agropecuary Research Corporation (AGROSAVIA), C.I. La Selva, Rionegro, Colombia

3 International Research Center in Agronomy for Development (CIRAD), Montpellier, France

*Address all correspondence to: fabrice.vaillant@cirad.fr

References

[1] Yoshida T, Amakura Y, Yoshimura M. Structural features and biological properties of ellagitannins in some plant families of the order Myrtales. International Journal of Molecular Sciences. 2010;**11**:79-106

[2] Yamada H, Wakamori S. Structural revisions in natural ellagitannins. Molecules. 2018;**23**:1-46

[3] Smeriglio A, Barreca D, Bellocco E, Trombetta D. Proanthocyanidins and hydrolysable tannins: Occurrence, dietary intake and pharmacological effects. British Journal of Pharmacology. 2017;**174**(11):1244-1262

[4] Alezandro MR, Dubé P, Desjardins Y, Lajolo FM, Genovese MI. Comparative study of chemical and phenolic composition of two species of Jaboticaba: *Myrciaria jaboticaba* (Vell.) berg and *Myrciaria cauliflora* (Mart.) O. Berg. Food Research International. 2013;**54**(1):468-477. DOI: 10.1016/j.foodres.2013.07.018

[5] Rojas-garbanzo C, Winter J, Montero ML, Zimmermann BF, Schieber A. Characterization of phytochemicals in Costa Rican guava (*Psidium friedrichsthalianum*-Nied.) fruit and stability of main compounds during juice processing—(U)HPLC-DAD-ESI-TQD-MSn. Journal of Food Composition and Analysis. 2019;**75**:26-42. DOI: 10.1016/j.jfca.2018.09.012

[6] Bakkalbasi E, Mentes O, Artik N. Food ellagitannins Occurrence, effects of processing and storage. Critical Reviews in Food Science and Nutrition. 2009;**49**(3):283-298

[7] Lipińska L, Klewicka E, Sójka M. Structure, occurrence and biological activity of ellagitannins: A general review. Acta Scientiarum Polonorum. Technologia Alimentaria. 2014;**13**(3):289-299

[8] Lorenzo M, Munekata PE, Putnik P. Sources, chemistry, and biological potential of ellagitannins and ellagic acid derivatives. In: Studies in Natural Products Chemistry. Elsevier Inc; 2018. pp 189-221. DOI: 10.1016/B978-0-444-64181-6.00006-1

[9] Bresciani L, Lean MEJ, Borges G, Calani L, Pereira-Caro G, Crozier A, et al. New insights into the bioavailability of red raspberry anthocyanins and ellagitannins. Free Radical Biology & Medicine. 2015;**89**:758-769. DOI: 10.1016/j.freeradbiomed.2015.10.400

[10] Koponen JM, Happonen AM, Mattila PH, Torronen AR. Contents of anthocyanins and ellagitannins in selected foods consumed in Finland. Journal of Agricultural and Food Chemistry. 2007;**55**:1612-1619

[11] Roquim M, Granato D, Inés M. Jaboticaba (*Myrciaria jaboticaba* (Vell.) Berg), a Brazilian grape-like fruit, improves plasma lipid profile in streptozotocin-mediated oxidative stress in diabetic rats. Food Reasearch International. 2013;**54**(1):650-659. DOI: 10.1016/j.foodres.2013.07.041

[12] Teixeira LL. Potential antiproliferative activity of polyphenol metabolites against human breast cancer cells and their urine excretion pattern in healthy subjects following acute intake of a polyphenol-rich juice of grumixama (*Eugenia brasiliensis* Lam.). Food & Function. 2017;**8**(6):2266-2274

[13] Abe LT, Lajolo FM, Genovese MI. Comparison of phenol content and antioxidant capacity of nuts. Food Science and Technology. 2010;**30**(1):254-259

[14] González-Sarrías A, Giménez-Bastida JA, García-Conesa MT, Gómez-Sánchez MB, García-Talavera NV,

Gil-Izquierdo A, et al. Occurrence of urolithins, gut microbiota ellagic acid metabolites and proliferation markers expression response in the human prostate gland upon consumption of walnuts and pomegranate juice. Molecular Nutrition & Food Research. 2010;**54**:311-322

[15] Adamczyk B, Simon J, Kitunen V, Adamczyk S. Tannins and their complex interaction with different organic nitrogen compounds and enzymes: Old paradigms versus recent advances. ChemistryOpen. 2017:610-614

[16] Clifford MN, Scalbert A. Ellagitannins—Nature, occurrence and dietary burden. Journal of the Science of Food and Agriculture. 2000;**80**:1118-1125

[17] Quideau S, Feldman KS. Ellagitannin chemistry. Chemical Reviews. 1996;**96**:475-503

[18] Aguilera-Carbo A, Augur C, Prado-Barragan LA, Favela-Torres E, Aguilar CN. Microbial production of ellagic acid and biodegradation of ellagitannins. Applied Microbiology and Biotechnology. 2008;**78**:189-199

[19] Larrosa M, Tomás-Barberán FA, Espín JC. The dietary hydrolysable tannin punicalagin releases ellagic acid that induces apoptosis in human colon adenocarcinoma Caco-2 cells by using the mitochondrial pathway. The Journal of Nutritional Biochemistry. 2006;**17**(9):611-625

[20] Okuda T, Ito H. Tannins of constant structure in medicinal and food plants—Hydrolyzable tannins and polyphenols related to tannins. Molecules. 2011:2191-2217

[21] Landete JM. Ellagitannins, ellagic acid and their derived metabolites: A review about source, metabolism, functions and health. Food Reasearch International. 2011;**44**(5):1150-1160. DOI: 10.1016/j.foodres.2011.04.027

[22] Khanbabaee K, Van RT. Tannins: Classification and definition. The Royal Society of Chemistry. 2001;**18**:641-649

[23] Sójka A, Karlińska E, Klewicki R. Ellagitannin and anthocyanin retention in osmotically dehydrated blackberries. Food Science and Technology Research. 2017;**23**(6):801-810. Available from: https://www.jstage.jst.go.jp/article/fstr/23/6/23_801/_article

[24] Hager TJ, Howard LR, Prior RL. Processing and storage effects on the ellagitannin composition of processed blackberry products. Journal of Agricultural and Food Chemistry. 2010;**58**(22):11749-11754

[25] Gancel AL, Feneuil A, Acosta O, Pérez AM, Vaillant F. Impact of industrial processing and storage on major polyphenols and the antioxidant capacity of tropical highland blackberry (*Rubus adenotrichus*). Food Research International. 2011;**44**(7):2243-2251. DOI: 10.1016/j.foodres.2010.06.013

[26] Mazur SP, Nes A, Wold A, Remberg SF, Martinsen BK, Aaby K. Effect of genotype and storage time on stability of colour, phenolic compounds and ascorbic acid in red raspberry (*Rubus idaeus* L.) jams. Acta Agriculturae Scandinavica. 2014;**64**(5):442-453

[27] Soto M, Acosta O, Vaillant F, Pérez A. Effects of mechanical and enzymatic pretreatments on extraction of polyphenols from blackberry fruits. Journal of Food Engineering. 2015;**39**(5):492-500

[28] Sebestyén Z, Jakab E, Badea E, Barta-Rajnai E, Şendrea C, Czégény Z. Thermal degradation study of vegetable tannins and vegetable tanned leathers. Journal of Analytical and Applied Pyrolysis. 2019;**138**:178-187. DOI: 10.1016/j.jaap.2018.12.022

[29] Garcia-Muñoz C, Vaillant F. Metabolic fate of ellagitannins: Implications for health, and research perspectives for innovative functional foods. Critical Reviews in Food Science and Nutrition. 2014;**54**(12):1584-1598

[30] Mena P, Calani L, Bruni R, Del RD. Bioactivation of high-molecular-weight polyphenols by the gut microbiome. In: Diet-Microbe Interactions in the Gut. Elsevier Inc; 2015. pp. 73-102. DOI: 10.1016/B978-0-12-407825-3.00006-X

[31] Sandhu AK, Miller MG, Thangthaeng N, Scott TM, Shukitt-Hale B, Edirisinghe I, et al. Metabolic fate of strawberry polyphenols after chronic intake in healthy older adults. Food & Function. 2018;**9**(1):96-106

[32] Andrade MA, Lima V, Sanches A, Vilarinho F, Conceição M, Khwaldia K, et al. Pomegranate and grape by-products and their active compounds: Are they a valuable source for food applications? Trends in Food Science & Technology. 2019;**86**(January):68-84. DOI: 10.1016/j.tifs.2019.02.010

[33] Nuñez-Sánchez MA, García-Villalba R, Monedero Saiz T, García-Talavera NV, Gómez-Sánchez MB, Sánchez-Álvarez C, et al. Targeted metabolic profiling of pomegranate polyphenols and urolithins in plasma, urine and colon tissues from colorectal cancer patients. Molecular Nutrition & Food Research. 2014;**58**(6):1199-1211

[34] Serrano J, Puupponen-pimiä R, Dauer A, Aura AM. Tannins: Current knowledge of food sources, intake, bioavailability and biological effects. Molecular Nutrition & Food Research. 2009;**53**:S310-S329

[35] Aguilar-Zárate P, Wong-Paz JE, Buenrostro-Figueroa JJ, Ascacio JA, Contreras-Esquivel JC, Aguilar CN. Ellagitannins: Bioavailability, purification and biotechnological degradation. Mini Reviews in Medicinal Chemistry. 2018;**18**(15):1244-1252

[36] Zhang X, Sandhu A, Edirisinghe I, Burton-Freeman B. An exploratory study of red raspberry (*Rubus idaeus* L.) (poly)phenols/metabolites in human biological samples. Food & Function. 2018;**9**(2):806-818

[37] Milala J, Kosmala M, Fotschki B. Ellagitannins from strawberries with different degrees of polymerization showed different metabolism through gastrointestinal tract of rats. Journal of Agricultural and Food Chemistry. 2017;**65**:10738-10748

[38] Liberal J, Carmo A, Gomes C, Cruz MT, Batista MT. Urolithins impair cell proliferation, arrest the cell cycle and induce apoptosis in UMUC3 bladder cancer cells. Investigational New Drugs. 2017;**35**(6):671-681

[39] García-Villalba R, Beltrán D, Espín JC, Selma MV, Tomás-Barberán FA. Time course production of urolithins from ellagic acid by human gut microbiota. Journal of Agricultural and Food Chemistry. 2013;**61**(37):8797-8806

[40] Truchado P, Larrosa M, Cerda B, Toma FA. Strawberry processing does not affect the production and urinary excretion of urolithins, ellagic acid metabolites, in humans. Journal of Agricultural and Food Chemistry. 2012;**60**(23):5749-5754

[41] Seeram NP, Henning SM, Zhang Y, Suchard M, Li Z, Heber D. Pomegranate juice ellagitannin metabolites are present in human plasma and some persist in urine for up to 48 hours. The Journal of Nutrition. 2006;**136**:2481-2485

[42] Cerdá B, Espín JC, Parra S, Martínez P, Tomás Barberán FA. The potent in vitro antioxidant ellagitannins from pomegranate juice are metabolised into bioavailable but poor antioxidant

hydroxy-6H-dibenzopyran-6-one derivatives by the colonic microflora of healthy humans. European Journal of Nutrition. 2004;**43**(4):205-220

[43] Mulero J, Tomás-barberán FA, Carlos J. Identifying the limits for ellagic acid bioavailability: A crossover pharmacokinetic study in healthy volunteers after consumption of pomegranate extracts. Journal of Functional Foods. 2015;**19**:225-235. DOI: 10.1016/j.jff.2015.09.019

[44] Gu J, Thomas-Ahner JM, Riedl KM, Bailey MT, Vodovotz Y, Schwartz SJ, et al. Dietary black raspberries impact the colonic microbiome and phytochemical metabolites in mice. Molecular Nutrition & Food Research. 2019;**63**(8):1800636. DOI: 10.1002/mnfr.201800636

[45] Garcia-Villalba R, Vissenaekens H, Pitart J, Romo-Vaquero M, Espín JC, Grootaert C, et al. The gastrointestinal simulation model TWIN-SHIME® shows differences between human urolithin-metabotypes in gut microbiota composition, pomegranate polyphenol metabolism, and transport along the intestinal tract. Journal of Agricultural and Food Chemistry. 2017;**65**(27):5480-5493

[46] Chambers K, Id EC, Pinto P, Ristic AK, Hollands WJ, Kroon PA, et al. Meta-analysis of the effects of foods and derived products containing ellagitannins and anthocyanins on cardiometabolic biomarkers: Analysis of factors influencing variability of the individual responses. International Journal of Molecular Sciences. 2018

[47] Cortés-Martín A, García-Villalba R, González-Sarrías A, Romo-Vaquero M, Loria-Kohenb V, Ramírez-de-Molina A, et al. The gut microbiota urolithin metabotypes revisited: The human metabolism of ellagic acid is mainly determined by aging. Food & Function. 2018;**9**(8):4100-4106

[48] González-Sarrías A, García-Villalba R, Romo-Vaquero M, Alasalvar C, Orem A, Zafrilla P, et al. Clustering according to urolithin metabotype explains the interindividual variability in the improvement of cardiovascular risk biomarkers in overweight-obese individuals consuming pomegranate: A randomised clinical trial. Molecular Nutrition & Food Research. 2017;**61**(5):1600830

[49] Romo-Vaqueroa M, Cortés-Martína A, Loria-Kohenb V, Ramírez-de-Molina A, García-Mantrana I, Colladoc MC, et al. Deciphering the human gut microbiome of urolithin metabotypes: Association with enterotypes and potential cardiometabolic health. Molecular Nutrition & Food Research. 2019;**63**(4):e1800958

[50] Selma MV, Beltrán D, Luna MC, Romo-Vaquero M, García-Villalba R, Mira A, et al. Isolation of human intestinal bacteria capable of producing the bioactive metabolite isourolithin A from ellagic acid. Frontiers in Microbiology. 2017;**8**:1521

[51] Romo-vaquero M, Espín JC, Tom FA, Selma MV. *Ellagibacter isourolithinifaciens* gen. nov., sp. nov., a new member of the family Eggerthellaceae, isolated from human gut. International Journal of Systematic and Evolutionary Microbiology. 2018;**68**(5):1707-1712

[52] Cerdá B, Tómas-Barberán FA, Espín JC. Metabolism of antioxidant and chemopreventive ellagitannins from strawberries, raspberries, walnuts, and oak-aged wine in humans: Identification of biomarkers and individual variability. Journal of Agricultural and Food Chemistry. 2005;**53**:227-235

[53] Seeram NP, Zhang Y, Mckeever R, Henning SM, Lee R, Suchard MA, et al. Ellagitannin metabolites in human subjects. Journal of Medicinal Food. 2008;**11**(2):390-394

[54] Cortés-Martín A, Selma MV, Espín JC, García-Villalba R. The human metabolism of nuts proanthocyanidins does not reveal urinary metabolites consistent with distinctive gut microbiota metabotypes. Molecular Nutrition & Food Research. 2019;**63**(2):e1800-e1819

[55] Derosa G, Maffioli P, Sahebkar A. Anti-inflammatory nutraceuticals and chronic diseases. In: Advances in Experimental Medicine and Biology. 2016. pp. 473-479. Available from: http://link.springer.com/10.1007/978-3-319-41334-1

[56] Umesalma S, Sudhandiran G. Differential inhibitory effects of the polyphenol ellagic acid on inflammatory mediators NF-κB, iNOS, COX-2, TNF-α, and IL-6 in 1,2-dimethylhydrazine-induced rat colon carcinogenesis. Basic & Clinical Pharmacology & Toxicology. 2010;**107**(2):650-655

[57] Ahire V, Mishra KP, Kulkarni G. Ellagic acid: A potent radio-sensitizer in cancer radiotherapy. Cancer Research Frontiers. 2016;**2**(2):141-155

[58] García-Muñoz C, Hernández L, Pérez A, Vaillant F. Diversity of urinary excretion patterns of main ellagitannins' colonic metabolites after ingestion of tropical highland blackberry (*Rubus adenotrichus*) juice. Food Research International. 2014;**55**:161-169. DOI: 10.1016/j.foodres.2013.10.049

[59] Lee G, Park J, Lee E, Ahn J, Kim H. Anti-inflammatory and antioxidant mechanisms of urolithin B in activated microglia. Phytomedicine. 2019;**55**:50-57. DOI: 10.1016/j.phymed.2018.06.032

[60] Singh R, Chandrashekharappa S, Bodduluri SR, Baby BV, Hegde B, Kotla NG, et al. Microbial metabolite through the Nrf2 pathway. Nature Communications. 2019;**10**(1):1-18. DOI: 10.1038/s41467-018-07859-7

[61] Boakye YD, Groyer L, Heiss EH. An increased autophagic flux contributes to the anti-inflammatory potential of urolithin A in macrophages. Biochimica et Biophysica Acta. 2017;**1862**(1):61-70. DOI: 10.1016/j.bbagen.2017.10.006

[62] Piwowarski JP, Stanisławska I, Granica S, Stefańska J, Kiss AK. Phase II conjugates of urolithins isolated from human urine and potential role of β-glucuronidases in their disposition. Drug Metabolism and Disposition. 2017;**45**(6):657-665

[63] Kiss AK, Piwowarski JP. Ellagitannins, gallotannins and their metabolites—The contribution to the anti-inflammatory effect of food products and medicinal plants. Current Medicinal Chemistry. 2018;**25**(37):4946-4967

[64] Dahiya NR, Chandrasekaran B, Kolluru V, Ankem M, Damodaran C, Vadhanam MV. A natural molecule, urolithin A, downregulates androgen receptor activation and suppresses growth of prostate cancer. Molecular Carcinogenesis. 2018;**57**(10):1332-1341

[65] Granica S, Piwowarski JP, Czarnocki Z, Kiss AK. The activity of urolithin A and M4 valerolactone, colonic microbiota metabolites of polyphenols, in a prostate cancer in vitro model. Planta Medica. 2019:118-125

[66] Zhou B, Wang J, Zheng G, Qiu Z. Methylated urolithin A, the modified ellagitannin-derived metabolite, suppresses cell viability of DU145 human prostate cancer cells via targeting miR-21. Food and Chemical Toxicology. 2016;**97**:375-384. DOI: 10.1016/j.fct.2016.10.005

[67] Paller CJ, Pantuck A, Carducci MA. A review of pomegranate in prostate cancer. Prostate Cancer and Prostatic Diseases. 2017;**20**(3):265-270. DOI: 10.1038/pcan.2017.19

[68] Paller CJ, Zhou XC, Heath EI, Taplin M, Mayer TM, Stein MN, et al. Muscadine grape skin extract (MPX) in men with biochemically recurrent prostate cancer: A randomized, multicenter, placebo-controlled clinical trial. Clinical Cancer Research. 2017;**24**(2):306-315

[69] Ávila-Gálvez MÁ, García-Villalba R, Martínez-Díaz F, Ocaña B, Monedero-Saiz T, Torrecillas-Sánchez A, et al. Metabolic profiling of dietary polyphenols and methylxanthines in normal and malignant mammary tissues from breast cancer patients. Molecular Nutrition & Food Research. 2019;Jan;**28**:e1801239

[70] Nuñez-Sánchez MA, González-Sarrías A, García-Villalba R, Monedero-Saiz T, García-Talavera NV, Gómez-Sánchez MB, et al. Gene expression changes in colon tissues from colorectal cancer patients following the intake of an ellagitannin-containing pomegranate extract: A randomized clinical trial. The Journal of Nutritional Biochemistry. 2017;**42**:126-133. DOI: 10.1016/j.jnutbio.2017.01.014

[71] Núñez-Sánchez MÁ, Karmokar A, González-Sarrías A, García-Villalba R, Tomás-Barberán FA, García-Conesa MT, et al. In vivo relevant mixed urolithins and ellagic acid inhibit phenotypic and molecular colon cancer stem cell features: A new potentiality for ellagitannin metabolites against cancer. Food and Chemical Toxicology. 2016;**92**:8-16. DOI: 10.1016/j.fct.2016.03.011

[72] Cho H, Jung H, Lee H, Yi C, Kwak H, Hwang KT. Chemopreventive activity of ellagitannins and their derivatives from black raspberry seeds on HT-29 colon cancer cells. Food & Function. 2015;**6**(5):1675-1683

[73] Nuñez-Sánchez MA, Dávalos A, González-Sarrías A, Casas-Agustench P, Visioli F, Monedero-Saiz T, et al. MicroRNAs expression in normal and malignant colon tissues as biomarkers of colorectal cancer and in response to pomegranate extracts consumption: Critical issues to discern between modulatory effects and. Molecular Nutrition & Food Research. 2015;**59**:1973-1986

[74] Selma MV, González-Sarrías A, Salas-Salvadó J, Andrés C, Alasalvar C, Örem A, et al. The gut microbiota metabolism of pomegranate or walnut ellagitannins yields two urolithin-metabotypes that correlate with cardiometabolic risk biomarkers: Comparison between normoweight, overweight-obesity and metabolic syndrome. Clinical Nutrition. 2017;**37**(3):897-905. DOI: 10.1016/j.clnu.2017.03.012

[75] Heiss C, Rodriguez-Mateos A, Istas G, Feliciano RP, Garcia-Villalba R, Weber T, et al. Plasma urolithin metabolites correlate with improvements in endothelial function after red raspberry consumption: A double-blind randomized controlled trial. Archives of Biochemistry and Biophysics. 2018;**651**:43-51. DOI: 10.1016/j.abb.2018.05.016

[76] Brighenti F, Bernini F, López-Gutiérrez N, Mena P, Marino V, Piemontese A, et al. Antiatherogenic effects of ellagic acid and urolithins in vitro. Archives of Biochemistry and Biophysics. 2016;**599**:42-50

[77] Han QA, Yan C, Wang L, Li G, Xu Y, Xia X. Urolithin A attenuates ox-LDL-induced endothelial dysfunction partly by modulating microRNA-27 and ERK/PPAR-γ pathway. Molecular Nutrition & Food Research. 2016;**60**(9):1933-1943

Chapter 3

Tannins as Antiviral Agents

Neli Vilhelmova-Ilieva, Angel S. Galabov and Milka Mileva

Abstract

Tannins possess a variety of biological effects, not a small part of which is of medical significance. Tannins, isolated from plants as well as synthetically obtained, manifest activity against a large spectrum of viruses: enteroviruses (polio- and coxsackie-), caliciviruses (feline calicivirus, mouse norovirus), rotavirus, influenza virus A, rhabdo- (vesicular stomatitis virus), paramyxoviruses (Sendai and Newcastle disease viruses), human immunodeficiency virus, herpes simplex virus, and adenoviruses. A special importance merits several ellagitannins manifesting pronounced effects against herpes simplex virus type 1 and 2 and on some herpes viruses affecting domestic animals, causing diseases of economic importance. An advantage of ellagitannins as anti-herpesvirus agents is that they have a non-nucleoside structure. Their targets are virus-specific proteins, so they retain activity against acyclovir-resistant strains of HSV types 1 and 2. Besides, these tannins manifest a synergistic effect with acyclovir when combined. Some initial results on their mechanism of action were carried out. In addition, it was found that most of the tannins have antioxidant properties in experimental models in vitro as well as in experimental influenza viral infection in mice.

Keywords: tannins, antiviral effect, antioxidant, herpes simplex virus, influenza virus

1. Introduction: tannins in medicine

For thousands of years throughout the world, tannins have been used in traditional medicines for the treatment of various health problems. They are used in the form of tea, coffee, and various extracts and in the daily intake of tannin-rich foods. They are also purposefully included as a component of many diets. Tannins are found in all parts of the plant—roots, stems, leaves, fruits, and seeds—which contribute to the existence of numerous natural sources of these substances.

Due to tannins' astringent effects, herbs containing tannins are used to treat injured and inflamed tissues, including burns, and to stop bleeding and prevent infection. Herbs contain tannins. They are also used in the treatment of mouth and throat inflammation, gastritis, enteritis, irritating bowel disorders, and other conditions [1–3], due to their ability to bind very tightly to proteins by forming multiple hydrogen bonds between their phenolic groups and the -NH groups of the peptides. Precipitation or the shrinking of proteins—the so-called tanning effect—then occurs, forming a protective layer on the surface of tissues.

Tannins have the ability to bind to metal ions in the body to form stable compounds called tannates. This property may have both positive and negative significance for human health. On the one hand, tannins can be used as an antidote for heavy metal poisoning. On the other hand, taking tannins daily, for example, in tea or coffee, can cause calcium and iron deficiencies in the body and may cause osteoporosis and anemia [4].

Tannins are also effective inhibitors of certain enzymes. For example, woodfruticosin (woodfordin C) shows anti-topoisomerase II activity; and eugeniflorin D1 and D2, isolated from *Eugenia uniflora* L., and oenothein B effectively inhibit Epstein-Barr virus (EBV) DNA polymerase. Oenothein A and B, isolated from *Epilobium* species, appear to be potential inhibitors of the enzymes 5α-reductase and aromatase, which play important roles in the development of benign prostatic hyperplasia. Researchers have suggested that the enzyme poly(ADP-ribose) glycohydrolase, an important factor in gene expression, DNA replication, and cell differentiation, can be inhibited by the oligomeric ellagitannins oenothein B and nobotanins B, E, and K. In addition, enzyme α-glucosidase (maltase), which may be important in the development of type-2 diabetes, is inhibited by chebulagic acid (isolated from *Terminalia chebula*), tellimagrandin I, and eugeniin (casuarictin) [5].

Research results indicate that the crude extract of *Terminalia bellerica* fruit (Tb. Cr), which is rich in tannin content, induced a dose-dependent fall in the arterial blood pressure of rats. Tb.Cr inhibited the force and rate of atrial contractions, and this effect was due to a calcium antagonistic mechanism [6]. Tannins also exhibit antihypertensive activity in the body by inhibiting the effect of angiotensin I-converting enzyme (ACE) [7]. Hydrolysable tannins with pronounced antihypertensive activity are castalagin and chebulinic acid as well as corilagin isolated from the leaves of *Lumnitzera racemosa* [8].

Important for human health is the antitumor activity different tannins show toward various tumors such as cervical and prostate cancer and malignant cells in the skin, breast, stomach, lung, esophagus, liver, and so on [9, 10], with several possible mechanisms of action. Ellagitannins possess the ability to bind to proteins located on the surface of the cell membrane, thus preventing the proliferation of metastatic cells. They can induce also apoptosis in tumor cells by inhibiting factors responsible for the formation of metastases. Another mechanism suggests that during DNA replication, ellagitannins bind carcinogens into a complex, so they cannot cause mutation [11]. There is also evidence that ellagitannins reduce the negative effects of chemotherapy and radiation in cancer treatment [12].

Tannins also show antimicrobial activity, in both plants and animals. In plants, the effect is due to the inhibition of microbial enzymes that degrade the plant cell wall [13]. Inhibition also occurs with other microbial enzymes such as pectinase, xylanase, peroxidase, laccase, and glycosyl transferase. Two possible mechanisms of this antimicrobial activity are tannins binding to the proteins of microbial membranes, damaging their structure and function, and tannins binding with essential metal ions [14]. Due to their antimicrobial activity, tannins can be used in the production and storage of certain foods to increase the shelf life of products [15].

The immunomodulatory activity of tannins has also been demonstrated, with different substances showing different mechanisms of action that enhance the functionality of macrophages [16–18] or stimulate the secretion of cytokines IL-1, IL-β2, and αTNF [18, 19].

2. Tannins as antiviral antioxidants

The overproduction of free radicals and the subsequent development of oxidative stress are implicated as pathogenic factors in a number of viral infections. Oxidative stress is a complex multifaceted biochemical condition, which occurs when there is an increase in oxidative damage to biomolecules and oxidation of nonprotein and protein thiols that regulate a cell's oxidative balance [20].

The cellular injury due to viral diseases caused by over generation of free radicals has been linked to over 200 clinical disorders [21].

It has been clearly established that many of viral infections trigger the production of reactive oxygen (ROS) and nitrogen (RNS) species. This is particularly true for infections caused by the blood-borne hepatitis viruses (B, C and D), human immunodeficiency virus (HIV), influenza virus, herpes simplex virus, Epstein-Barr virus, respiratory syncytial virus, coxsackievirus B3 (CVB3), and others. For acute respiratory viral infections, ROS/RNS have been implicated in lung tissue injury and epithelial barrier dysfunction, which in turn increased susceptibility to secondary infections [22].

A variety of DNA viruses are associated with the increased oxidative stress that promotes DNA damage, high mutagenicity, and initiation and/or progression of neoplasia [23].

Phenolic compounds such as phenolic acids, flavonoids, tannins, and proanthocyanidins are widely distributed in plants and are a protective mechanism against OS. Several studies, including in vivo (in experimental animals) and epidemiological investigations, have demonstrated that phenolic compounds in foods possess positive attributes such as antioxidant potential, which is the basis of antiviral, antimicrobial, and antimutagenic activities. Compounds present in food that have potential antioxidant activity include vitamins C, E, and K, phenols (phenolic acids, flavonoids, thymol, carvacrol, and tannins), and carotenoids [24, 25]. Thus, antioxidants, mainly those originating from natural products, are of great importance for human health.

Antioxidant therapy is becoming an attractive and effective alternative approach for the treatment of viral diseases. The antioxidant properties of apple polyphenol extract, which is rich in tannins, are effective against the development of influenza virus infection in mice—they improve survival rates and also significantly decrease lipid peroxidation and increase oxygen radical absorbance capacity (ORAC) in splenocytes [26].

Avian influenza is usually accompanied by virus invasion followed by the occurrence of oxidative stress and serious inflammation. The anti-inflammatory and antioxidant properties of tannin-rich extracts of *Chaenomeles speciosa* showed that the multiple effects of the isolates might play a cocktail-like role in the treatment of avian influenza, and *C. speciosa* components might be a potent source for antiviral and anti-inflammatory agents [27].

Pomegranate juice consumption reduces oxidative stress during influenza infection [28].

Green tea is an important source of polyphenol antioxidants, which are also rich in tannins. Tea polyphenols possess antiviral properties believed to help protect against influenza virus. Oxidative stress and inflammation in the oral cavity, due to cigarette smoking and cigarettes' deleterious compounds nicotine and acrolein, can be reduced by green tea polyphenols. Generally, green tea defends healthy cells from malignant transformation, and locally, it has the ability to induce apoptosis in oral cancer cells [29].

Finally, oxidative stress and the stress-mediated complications of viral infections successfully respond to antioxidant prevention. However, it should be kept in mind that antioxidants are not antivirals. Their function is more auxiliary, and they are particularly beneficial when used in combination therapy with specific viral inhibitors.

3. Tannins as antivirals

For a small fraction of today's known viral diseases, there are vaccines that can be successfully applied. Medicaments that are used are also limited in number, and in most cases, their use is accompanied by the appearance of side effects or the formation of resistant viral mutants, making therapy ineffective. Therefore, turning to nature to find effective therapies is a good solution to this problem. As mentioned earlier, tannins are a component of many plants. They are found in relatively high concentrations and exhibit significant biological activities. Tannin attack targets can carry out different stages of viral replication, including the extracellular virions themselves, their attachment to the cell, their penetration into the cell and the replication process in the host cell, as well as the assembling of new viral particles, transport proteins, polysaccharides, and viral enzymes [30, 31]. In almost all of the abovementioned stages, the tannin activity is due to their ability to bind permanently to the proteins of the capsid or supercapside, either to specific viral enzymes required for viral replication or to newly synthesized viral proteins involved in the composition of the new viral particles.

Numerous plant extracts have been studied, in which tannins are the main component, and they have shown good results against the replication of different viruses. The resulting effects have been on both coated viruses (influenza viruses A/H3N2 and A/H5N3, herpes simplex virus type 1 (HSV-1), vesicular stomatitis virus, Sendai virus and Newcastle disease viruses) [32, 33] and non-enveloped viruses (poliovirus, coxsackievirus, adenovirus, rotavirus, feline calicivirus, and mouse norovirus) [34].

The antiretroviral activity of *Euphorbia hirta* extracts with high tannin content has indicated a dose-dependent inhibition of reverse transcriptase activity in vitro [35].

Extracts of *Hamamelis virginiana* L. bark, with differing concentrations of tannins and individual tannins of defined structures, including pseudotannins, have been tested for effect against influenza A virus (IAV) and human papillomavirus (HPV) type 16 infections. The study demonstrated that the IAV life cycle is inhibited in the early and, to a minor extent, later steps and that HPV attachment is tannin-dependently inhibited. Of the investigated substances high molecular weight tannin inhibited both IAV receptor binding and neuraminidase activity. However, those with low molecular weight tannin inhibited neuraminidase but not hemagglutination [36].

Many tannins showing antiviral activity have been isolated and characterized. The hydrolyzable tannins chebulagic acid and punicalagin were identified as potent inhibitors of HCV entry [37]. The replication of human, porcine, and duck influenza A virus in vitro was prevented by the hydrolyzable tannin strictinin [38].

Finally, many studies have been conducted on tannins' effects against the replication of human immunodeficiency virus (HIV), and the results of the various teams indicate that tannins have several targets of action in the HIV replicative cycle. Ellagitannins isolated from *Tuberaria lignosa* inhibited HIV's entry into MT-2 cells [39]. There is evidence that ellagitannins suppressed HIV replication by inhibiting reverse transcriptase [40–44]. Other authors have reported on ellagitannins (geraniin and corilagin) that reduced HIV replication by inhibiting the HIV-1 protease and HIV-1 integrase enzymes [43].

4. Ellagitannins as antiherpesvirus agents

Various tannins have been tested for antiherpesviral activity. Ellagitannin geraniin possesses a virucidal effect against herpesviruses [45], and it inhibits the adsorption of HSV and HTLV-IIIB [46–49]. The hydrolyzable tannin casuarinin isolated from *Terminalia arjuna* Linn prevents the attachment of HSV-2 and its penetration into the cell, and it also disturbs the late stages of infection [1]. Chebulagic acid and punicalagin—two hydrolysable tannins isolated from *Terminalia chebula*

Retz.—inactivate HSV-1 entry and the cell-to-cell spread of the virus by targeting HSV-1 glycoproteins [50]. Putranjivain A isolated from *Euphorbia jolkini* inhibits the entry of the virus and the late stages of HSV-2 replication in vitro [51].

Seven ellagitannins isolated from *Phyllanthus myrtifolius* and *Phyllanthus urinary*, and eugeniflorin D (1) and D (2) isolated from *Eugenia uniflora* L., are active against the DNA polymerase of EBV [52, 53]. *Eucalyptus grandis* extract containing euglobal-G1 and euglobal-G3 shows antiviral activity against EBV, as do quassinoids (ailantinol B, ailantinol C, and ailanthone). Eugenol and eugenin isolated from *Geum japonicum* or *Syzygium aromaticum* show inactivating activity on viral DNA polymerase and thus inhibit acyclovir-resistant TK-deficient HSV-1 virus, wild HSV-2, and EBV [52, 54–56]. The EBV DNA polymerase is also inhibited by ellagitannins contained in *Phyllanthus myrtifolius* extracts and *Phyllanthus urinaria* (*Euphorbiaceae*), probably due to the corilagin moiety of these tannins. The tannin samarangenin B contained in the alcoholic extract of *Limonium sinensis* significantly suppresses HSV-1 multiplication [57]. And cowaniin isolated from *Cowania mexicana* (*Rosaceae*) exhibits an inhibitory effect on the activation of EBV early antigens [58].

Our studies on the antiviral activity of tannins have been mainly related to substances that belong to the group of nonahydroxyterphenoil-bearing C-glucosidic ellagitannins. We investigated the activity of three compounds—castalagin, vescalagin, and grandin isolated from powdered pedunculate oak (i.e., *Quercus robur*)—against the replication of two strains of HSV-1 (DA and Victoria) susceptible to acyclovir (ACV), one HSV-1 strain resistant to ACV (R-100), two HSV-2 strains susceptible to ACV (XA and Bja), and one HSV-2 strain resistant to ACV (PU) (**Table 1**) [59, 60].

Currently existing therapy against HSV infections is based on the administration of nucleoside analogues, among which ACV has had the broadest application. A disadvantage of this therapy is the rapid formation of resistant mutants [61].

All three investigated ellagitannins showed remarkable antiviral activity against all strains of HSV-1, the strongest being castalagin's action against the DA strain (SI = 5390.0) followed by vescalagin's action against that strain (SI = 4546.0), and these effects were greater than that of ACV. The effects on the replication of HSV-2 strains, although less pronounced than those on HSV-1, were significant. The strongest effects were seen on strain XA, with the following SI values: vescalagin = 378.9, castalagin = 336.9, and grandinin = 208.8.

The activity of the three ellagitannins was also determined in relation to the replication of three of the most common and important HSV-1 strains of economic importance, namely, pseudorabies virus or Suid herpesvirus 1 (SuHV-1), bovine

Compounds	SI = CC_{50}/IC_{50}					
	HSV-1			HSV-2		
	Strains sensitive to ACV		Strain resistant to ACV	Strains sensitive to ACV		Strain resistant to ACV
	DA	Victoria	R-100	XA	Bja	PU
Castalagin	5390.0	4498.0	1047.5	336.9	89.9	97.4
Vescalagin	4546.0	909.0	900.0	378.9	123.9	117.4
Grandinin	1183.0	88.7	650.0	208.8	71.0	103.1
ACV	1270.5	972.8	25.3	790.2	810.5	29.0

**The data for strains DA and DX are the original ones; the rest are from references [59, 60].*

Table 1.
*Effect of ellagitannins on HSV replication.**

herpesvirus-1 (BoHV-1), and caprine herpesvirus-1 (CapHV-1). The effect of the three ellagitannins was strongest against SuHV-1, with the activity of castalagin (SI = 336.8) and vescalagin (SI = 309) being on the order of ACV, while grandinin exhibited a moderate effect (SI = 40.8). The activity of the three ellagitannins against BoHV-1 was comparatively weaker but still significant (castalagin SI = 45, vescalagin SI = 42.5, grandinin SI = 32.3). Activity against the CapHV-1 strain had limited values.

Antiviral activity against HSV-1 (Victoria strain) was also determined for nine ellagitannins, of which six are natural compounds (castalin, vescalin, acutissimin A, epiacutissimins (EPI) A and B, mongolicain) and three are vescalagin synthetic derivatives (VgSBuSH, VgSOctSH, VgOMe). Thirteen gallotannin-type compounds [Gal-01A, Gal-01B, Gal-02A, Gal-02B, Gal-03 M, Gal-04A, Gal-04B, Gal-05 M, Gal-07, Gal-08, Gal-09, Gal-11 M (tannic acid), Gal-12 (gallic acid)] as well as Gal-13 and Gal-14 (ellagic acid)] were also tested. Generally, the group of ellagitannins exhibited greater activity, with only castalin and vescalin, from the natural products, and one of the synthetic derivatives (VgSOctSH) showing no activity. The remaining four natural components exhibited more pronounced activity than did the synthetic products, with the strongest effect showing for Epi B and Epi A (**Table 2**). Only three of the gallotannins—Gal-04A, Gal-04B, and Gal-11 M—showed activity against HSV-1 replication (**Table 2**) [62].

In order to control HSV infections, especially in immunosuppressed patients, it is necessary to treat them with antiherpetic preparations. However, their systemic use leads to the selection and/or formation of resistant strains. Given the heterogeneity of the viral population, naturally resistant variants are present. Therefore, the use of new approaches to the treatment of HSV infections [63, 64], namely, by combination therapy with two or more chemotherapeutics with synergistic action, is being sought. In these new approaches, synergistic combinations of two and more preparations are used to attack multiple viral targets simultaneously. This reduces the possibility of the formation and/or selection of resistant mutants. Even more, the therapeutic doses are abruptly reduced, eliminating any side effects. Data have been reported for combinations of antiviral agents showing a synergistic or additive effect on HSV [65–68].

Each of the three ellagitannins—castalagin, vescalagin, and grandinin—was administered in combination with ACV, and their effects against the replication

Compound	MM (g/mol)	CC_{50} (μM)	IC_{50} (μM)	SI = CC_{50}/IC_{50}
Epi A	1207	>1000***	18.0 ± 0.77***	>55.5
Epi B	1207	>1000***	16.5 ± 0.14***	>60.6
Acutissimin	1207	>640***	18.4 ± 1.2***	>34.78
Mongolicain	1177	>640***	19.7 ± 0.84***	>32.5
VgSBuSH	1039	>640***	26.0 ± 1.76***	>24.6
VgOMe	949	>640***	29.0 ± 2.89***	>22.0
Gal-04A	1701	>200***	7.0 ± 1.2**	>28.5
Gal-04B	1701	>100***	2.8 ± 0.53*	>35.7
Gal-11 M(TA)	1701	>100***	4.0 ± 0.07*	>25.0

The table presents data partially contained in Ref. [62]. *$p > 0.05$.*
**$p < 0.05$.*
***$p < 0.001$, *when comparing the value of each gallotannin against ACV (CC_{50} = 1296.0 μM; IC_{50} = 1.47 μM) [59].*

Table 2.
Effect of tannins on the replication of HSV-1 in MDBK cell.

of HSV-1 and HSV-2 strains sensitive and resistant to ACV's effect were markedly synergistic [59, 60]. To evaluate the effect of the combinations, we employed the three-dimensional model system developed by Prichard and Shipman [69] using the computer program MacSynergy ™ II [70]. The program calculates the volume of synergy in μM^2%, where values between 50 and 100 μM^2% indicate moderate synergy (this interaction may be important in vivo) and values over 100 μM^2% indicate strong synergy (these are more likely to be important in vivo). The strongest synergistic effect was seen in the combinations administered against the ACV-resistant HSV-1 strain, and the effect was also pronounced in the ACV-sensitive HSV-1 strain. The combined effect on the HSV-2 strains was weaker but also significant: in both the resistant and sensitive strains, the effect was on the same order of magnitude (**Table 3**).

The telling synergistic effect of all three ellagitannins shows that they have a different mechanism of action against HSV reproduction compared to that of ACV. The exact mechanism of antiviral activity of tannins has not been studied in detail.

To elucidate the mechanism of anti-herpes activity of tannins, we used a substance that showed activity similar to that of acyclovir, namely, castalagin.

When monitoring castalagin's effect on extracellular virions, we found that it was markedly time dependent. At the first time interval—15 min—the effect was negligible, but at 30 min, the effect was already significant, and as time increased, the virucidal effect intensified. This effect was also influenced by the temperature at which it occurred, with the effect at 37°C being stronger than that at room temperature [71].

Castalagin's effect on the attachment of HSV-1 virions to MDBK cells was time dependent, and it was also dependent on the concentration of castalagin and the number of infectious viral particles. The inhibitory effect was reported at 30 min (Δlog 1.7), and it increased with the time of impact, reaching a value of Δlog 3.2 at 60 min. The most remarkable effect was observed when using castalagin at a maximum non-toxic concentration of 10 μM and then reducing its effect with a decrease in its concentration [71].

Using a one-staged viral replicative cycle in a timing-of-addition study, we tested castalagin's effect on the production of infectious virions during the replication cycle of ACV-sensitive HSV-1, Victoria strain. The effect of adding castalagin at 0 hours was most pronounced, and the effect remained significant when adding the ellagitannin at up to 4 hours. After this period of time, the addition of the substance had no particular effect, and after 15 hours, a statistically significant effect was not observed. From these results, it can be concluded that castalagin affects earlier stages of the viral replicative cycle [71].

These results demonstrate a very high activity of the ellagitannin derivative castalagin toward human herpes simplex viruses 1 and 2. Its effect is on the order of the most efficient anti-HSV compound, acyclovir, which is widely used in clinical practice. In addition, castalagin manifested a marked activity against ACV-resistant

Compound	**HSV-1**		**HSV-2**	
	Victoria	**R-100**	**Bja**	**PU**
Castalagin	222.06	222.13	71.62	97.78
Vescalagin	205.09	324.44	87.56	79.11
Grandinin	106.0	314.53	132.78	106.12

*The table is originally constructed with results presented in Refs. [59, 60].

Table 3.
Synergistic effect between ellagitannins and acyclovir (μM^2%).

HSV strains, and its combination effects with ACV could be characterized as synergistic. Another advantage of this substance is its non-nucleoside chemical structure.

Castalagin could be considered as a candidate for in vivo testing on experimental HSV infections in laboratory animals, such as HSV-1-induced skin infection in mice, encephalitis in newborn mice, eye infection in rabbits, as well as HSV-2 genital infection in mice. A very important step in the characterization of the anti-HSV effect of castalagin is the determination of its target in the herpesvirus replication cycle via molecular genetic analysis.

Author details

Neli Vilhelmova-Ilieva, Angel S. Galabov* and Milka Mileva
Bulgarian Academy of Sciences, The Stephan Angeloff Institute of Microbiology, Sofia, Bulgaria

*Address all correspondence to: galabov@microbio.bas.bg

References

[1] Cheng HY, Lin CC, Lin TC. Antiherpes simplex virus type 2 activity of casuarinin from the bark of *Terminalia arjuna* Linn. Antiviral Research. 2002;**55**:447-455

[2] Ashok PK, Upadhyaya K. Tannins are Astringent. Journal of Pharmacognosy and Phytochemistry. 2012;**1**(3):45-50. ISSN 2278-4136

[3] Jaiswal H, Singh OJ, Chauhan A, Sahu MK, Prakash S. DV. A review on tannins. European Journal of Biotechnology and Bioscience. 2018;**6**(9):16-17. ISSN: 2321-9122

[4] Ricardo-da-Silva JM, Cheynier V, Souquet J, Moutounet M. Interaction of grape seed procyanidins with various proteins in relation to wine fining. Journal of the Science of Food and Agriculture. 1991;**57**:111-125

[5] Yoshida T, Amakura Y, Yoshimura M. Structural features and biological properties of ellagitannins in some plant families of the order Myrtales. International Journal of Molecular Sciences. 2010;**11**:79-106

[6] Khan A-U, Gilani AH. Pharmacodynamic evaluation of *Terminalia bellerica* for its antihypertensive effect. Journal of Food and Drug Analysis. 2008;**16**(3):6-14

[7] Liu J-C, Hsu F-L, Tsai J-C, Chan P, Liu JY-H, Thomas GN, et al. Antihypertensive effects of tannins isolated from traditional Chinese herbs as non-specific inhibitors of angiotensin converting enzyme. Life Sciences. 2003;**73**(12):1543-1555. DOI: 10.1016/S0024-3205(03)00481-8

[8] Lin TC, Hsu FL, Cheng JT. Antihypertensive activity of corilagin and chebulinic acid, tannins from *Lumnitzera racemosa*. Journal of Natural Products. 1993;**56**:629-632

[9] Ascacio-Valdés J, Buenrostro-Figueroa JJ, Aguilera-Carbo A, Prado-Barragán A, Rodríguez-Herrera R, Aguilar CN. Ellagitannins: Biosynthesis, biodegradation and biological properties. Journal of Medicinal Plants Research. 2011;**5**(19):4696-4703

[10] Yildirim I, Kutlu T. Anticancer agents: Saponin and tannn. International Journal of Biological Chemistry. 2015;**9**:332-340

[11] Sepúlveda L, Ascacio A, Rodriguez-Herrera R, Aguilera-Carbó A, Aguilar CN. Ellagic acid: Biological properties and biotechnological development for production processes. African Journal of Biotechnology. 2011;**10**(22):4518-4523

[12] Varadkar P, Dubey P. Modulation of radiation-induced protein kinase C activity by phenolics. Journal of Radiological Protection. 2001;**21**:361-370

[13] Heldt H-W, Piechulla B. Phenylpropanoids comprise a multitude of plant secondary metabolites and cell wall components. In: Plant Biochemistry. 4th ed. Amsterdam, Boston: Academic Press; 2011. pp. 431-449. DOI: 10.1016/B978-0-12-384986-1.00018-1

[14] Ribeiro M, Simões L, Simões M. Biocides. In: Reference Module in Life Sciences. Porto: ScienceDirect; 2018. DOI: 10.1016/B978-0-12-809633-8.12118-1

[15] Bajaj YS. Medicinal and Aromatic Plants V. Biotechnology in Agriculture and Forestry. Berlin, Heidelberg: Springer; 1999. p. 24

[16] Ushio Y, Fang T, Okuda T, Abe H. Modificational changes in function and morphology of

cultured macrophages by geraniin. Japanese Journal of Pharmacology. 1991;**57**:187-196

[17] Kolodzej H, Kiderlen AF. Antileishmanial activity and immune modulatory effects of tannins and related compounds on *Leishmania* parasitised RAW 264.7 cells. Phytochemistry. 2005;**66**(17):2056-2071

[18] Schepetkin IA, Kirpotina LN, Jakiw L, Khlebnikov AI, Blaskovich CL, Jutila MA, et al. Immunomodulatory activity of Oenothein B isolated from *Epilobium angustifolium*. Journal of Immunology. 2009;**183**(10):6754-6766. DOI: 10.4049/jimmunol.0901827

[19] Wang CC, Chen LG, Yang LL. *In vitro* immunomodulatory effects of cuphiin D 1 on human mononuclear cells. Anticancer Research. 2002;**22**:4233-4236

[20] Halliwell B, Gutteridge JMC. Cellular responses to oxidative stress: Adaptation, damage, repair, senescence and death. In: Halliwell B, Gutteridge JMC, editors. Free Radicals in Biology and Medicine. 4th ed. New York: Oxford University Press; 2007. pp. 187-267

[21] Gacche R, Khsirsagar M, Kamble S, Bandgar B, Dhole N, Shisode K, et al. Antioxidant and anti-inflammatory related activities of selected synthetic chalcones: Structure–activity relationship studies using computational tools. Chemical & Pharmaceutical Bulletin. 2008;**56**(7):897-901

[22] De Marco F. Oxidative stress and HPV carcinogenesis. Viruses. 2013;**5**(2):708-731

[23] Ivanov A, Bartosch B, Isaguliants M. Oxidative stress in infection and consequent disease. Oxidative Medicine and Cellular Longevity. 2017;**2017**:3496043

[24] Carocho M, Ferreira ICFR. A review on antioxidants, prooxidants and related controversy: Natural and synthetic compounds, screening and analysis methodologies and future perspectives. Food and Chemical Toxicology. 2013;**51**(1):15-25. View at Publisher View at Google Scholar View at Scopus

[25] Miguel M. Antioxidant activity of medicinal and aromatic plants. A review. Flavour and Fragrance Journal. 2010;**25**(5):291-312

[26] He RR, Wang M, Wang CZ, Chen BT, Lu CN, Yao XS, et al. Protective effect of apple polyphenols against stress-provoked influenza viral infection in restraint mice. Journal of Agricultural and Food Chemistry. 2011;**59**(8):3730-3737

[27] Zhang L, Cheng YX, Liu AL, Wang HD, Wang YL, Du GH. Antioxidant, anti-inflammatory and anti-influenza properties of components from Chaenomeles speciosa. Molecules. 2010;**15**(11):8507-8517

[28] Haidari M, Ali M, Casscells SW III, Madjid M. Pomegranate (Punica granatum) purified polyphenol extract inhibits influenza virus and has a synergistic effect with oseltamivir. Phytomedicine. 2009;**16**(12):1127-1136

[29] Narotzki B, Reznick AZ, Aizenbud D, Levy Y. Green tea: A promising natural product in oral health. Archives of Oral Biology. 2012;**57**(5):429-435

[30] Chattopadhyay D. Ethnomedicin al antivirals: Scope and opportunity. Chapter 15. In: Modern Phytomedicine: Turning Medicinal Plants into Drugs. CRC Press; 2006. pp. 313-338

[31] Chattopadhyay D, Naik TN. Antivirals of ethnomedicinal origin: Structure-activity relationship and

scope. Mini Reviews in Medicinal Chemistry. 2007;**7**(3):275-301. (Review)

[32] Uozaki M, Yamasaki H, Katsuyama Y, Higuchi M, Higuti T, Koyama AH. Antiviral effect of octyl gallate against DNA and RNA viruses. Antiviral Research. 2007;**73**:85-91

[33] Kratz JM, Andrighetti-Fröhner CR, Kolling DJ, Leal PC, Cirne-Santos CC, Yunes RA, et al. Anti-HSV-1 and anti-HIV-1 activity of gallic acid and pentyl gallate. Memórias do Instituto Oswaldo Cruz. 2008;**103**(5):437-442

[34] Ueda K, Kawabata R, Irie T, Nakai Y, Tohya Y, Sakaguchi T. Inactivation of pathogenic viruses by plant-derived tannins: Strong effects of extracts from persimmon (*Diospyros kaki*) on a broad range of viruses. PLoS One. 2013;**8**(1):e55343

[35] Gyuris Á, Szlávik L, Mináróvits J, Vasas A, Molnár J, Hohmann J. Antiviral activities of extracts of Euphorbia hirta L. against HIV-1, HIV-2 and SIV_{mac251}. In vivo. 2009;**23**(3):429-432

[36] Theisen LL, Erdelmeier CAJ, Spoden GA, Boukhallouk F, Sausy A, Florin L, et al. Tannins from *Hamamelis virginiana* bark extract: Characterization and improvement of the antiviral efficacy against influenza a virus and human papillomavirus. Plos One. 2014. DOI: 10.1371/journal.pone.0088062

[37] Lin LT, Chen TY, Lin SC, Chung CY, Lin TC, Wang GH, et al. Broad-spectrum antiviral activity of chebulagic acid and punicalagin against viruses that use glycosaminoglycans for entry. BMC Microbiology. 2013;**13**:187

[38] Saha RK, Takahashi T, Kurebayashi Y, Fukushima K, Minami A, Kinbara N, et al. Antiviral Research. 2010;**88**:10-18

[39] Bedoya LM, Abadb MJ, Sánchez-Palominoa S, Alcamia J, Bermejob P. Ellagitannins from Tuberaria lignosa as entry inhibitors of HIV. Phytomedicine. 2010;**17**:69-74

[40] Kakiuchi N, Hattori M, Namba T, Nishizawa M, Yamagishi T, Okuda T. Journal of Natural Products. 1985;**48**:614

[41] Matthée G, Wright AD, König GM. HIV reverse transcriptase inhibitors of natural origin. Planta Medica. 1999;**65**(6):493-506

[42] Asanaka M, Kurimura T, Kobayashi R, Okuda T, Mori M, Yokoi H. Fourth International Conference on Immunopharmacol. Osaka, Japan: Thieme; 1988. p. 47

[43] Notka F, Meier G, Wagner R. Concerted inhibitory activities of Phyllanthus amarus on HIV replication in vitro and ex vivo. Antiviral Research. 2004;**64**:93-102

[44] Notka F, Meier GR, Wagner R. Inhibition of wild-type human immunodeficiency virus and reverse transcriptase inhibitor-resistant variants by Phyllanthus amarus. Antiviral Research. 2003;**58**:175-186

[45] Yang C, Cheng H, Lin T, Chiang L, Lin C. The in vitro activity of geraniin and 1,3,4,6-tetra-O-galloyl-β-d-glucose isolated from Phyllanthus urinaria against herpes simplex virus type 1 and type 2 infection. Journal of Ethnopharmacology. 2007;**110**:555-558

[46] Li J, Huang H, Zhou W, Feng M, Zhou P. Anti-hepatitis B virus activities of *Geranium carolinianum* L. extracts and identification of the active compounds. Biological & Pharmaceutical Bulletin. 2008;**3**:743-747

[47] Fukuchi K, Sagakami H, Okuda T, Hatano T, Tanuma S, Katajima K, et al. Inhibition of herpes simplex virus infection by tannins and related compounds. Antiviral Research. 1989;**11**:285-297

[48] Okuda T, Yoshida T, Hatano T. In: Huang MT, Ho CT, Lee CY, editors. Phenolic Compounds in Food and their Effects on Health, II. Washington, DC: ACS Symposium Series 507: American Chemical Sosiety; 1992a. p. 87

[49] Okuda T, Yoshida T, Hatano T. In: Hemingway RW, Lacks PE, editors. Plant Polyphenols Synthesis, Properties, Significance. New York: Plenum Press; 1992b. p. 539

[50] Lin L-T, Chen T-Y, Chung C-Y, Noyce RS, Grindley TB, McCormick C, et al. Hydrolyzable tannins (chebulagic avid and punicalagin) target viral glycoprotein-glycosaminoglycan interactions to inhibit herpes simplex virus 1 entry and cell-to-cell spread. Journal of Virology. 2011;**85**:4386-4398

[51] Cheng HY, Lin TC, Yang CM, Wang KC, Lin LT, Lin CC. Putranjivain a from *Euphorbia jolkini* inhibits both virus entry and late stage replication of herpes simplex virus type 2 *in vitro*. The Journal of Antimicrobial Chemotherapy. 2004;**53**:577-583

[52] Lee MH, Chiou JF, Yen KY, Yang LL. EBV DNA polymerase inhibition of tannins from *Eugenia uniflora*. Cancer Letters. 2000;**154**:131-136

[53] Kim HJ, Lee J, Woo ER, Kim MK, Yang BS, Yu YG, et al. Isolation of virus-cell fusion inhibitory components from *Eugenia caryophyllata*. Planta Medica. 2001;**67**:277-279

[54] Kurokawa M, Hozumi T, Basnet P, Nakano M, Kadota S, Namba T, et al. Purification and characterization of eugeniin as an anti-herpesvirus compound from *Geum japonicum* and *Syzygium aromaticum*. The Journal of Pharmacology and Experimental Therapeutics. 1998;**284**:728-735

[55] Kurokawa M, Hozumi T, Tsurata M, Kadota SH, Namba T, Shiraki K. Biological characterization of eugeniin as an anti-herpes simplex virus type 1 compound *in vitro* and *in vivo*. The Journal of Pharmacology and Experimental Therapeutics. 2001;**297**:372-379

[56] Chattopadhyay D, Khan MTH. Ethnomedicines and ethnomedicinal phytophores against Herpesviruses. Biotechnology Annual Review. 2008;**14**:297-349. ISSN 1387-2656. DOI: 10.1016/S1387-2656(08)00012-4

[57] Chattopadhyay D, Das S, Chakrabory S, Bhattacharya SK. Ethnomedicines for the development of anti-herpesvirus agents. In: Ethnomedicine: A Source of Complementary Therapeutics. 2010. pp. 117-147

[58] Ito H. Metabolites of the ellagitannin geraniin and their antioxidant activities. Planta Medica. 2011;**77**:1110-1115

[59] Vilhelmova N, Jacquet R, Quideau S, Stoyanova A, Galabov AS. Three-dimensional analysis of combination effect of ellagitannins and acyclovir on herpes simplex virus types 1 and 2. Antiviral Research. 2011;**89**(2):174-181

[60] Vilhelmova-Ilieva N, Jacquet R, Quideau S, Galabov AS. Ellagit annins as synergists of ACV on the replication of ACV-resistant strains of HSV 1 and 2. Antiviral Research. 2014;**110**:104-114

[61] Abraham AM, Kavitha S, Joseph P, George R, Pillay D, Malathi J, et al. Aciclovir resistance among indian strains of herpes simplex virus as determined using a dye uptake assay. Indian Journal of Medical Microbiology. 2007;**25**:260-262

[62] Vilhelmova-Ilieva N, Jacquet R, Deffieux D, Pouységu L, Sylla T, Chassaing S, et al. Anti-herpes simplex virus type 1 activity of specially selected groups of tannins. Drug Research. 2018;**1-8**:68

[63] Pillay D, Mutimer D, Singhal S, Turner A, Ward K, Wood M. Management of herpes virus infections following transplantation. The Journal of Antimicrobial Chemotherapy. 2000;**45**:729-748

[64] Frangoul H, Wills M, Crossno C, Engel M, Domm J. Acyclovir-resistant herpes simplex virus pneumonia post-unrelated steam cell transplantation: A word of caution. Pediatric Transplantation. 2007;**11**:942-944

[65] Lerner AM, Bailey EJ. Differential sensitivity of herpes simplex virus types 1 and 2 to human interferon: Antiviral effects of interferon plus 9-β-D-arabinofuranosyladenine. The Journal of Infectious Diseases. 1976;**134**:400-404

[66] Schinazi RF, Peters J, Williams CC, Chance D, Nahmias AJ. Effect of combinations of acyclovir with vidarabine or its 5′-monophosphate on herpes simplex virus in cell culture and in mice. Antimicrobial Agents and Chemotherapy. 1982;**22**:499-507

[67] Piret J, Roy S, Gagnon M, et al. Comparative study of mechanisms of herpes simplex virus inactivation by sodium lauryl sulfate and n-lauroylsarcosine. Antimicrobial Agents and Chemotherapy. 2002;**46**:2933-2942

[68] Kawaguchi K, Inamura H, Abe Y, Koshu H, Takashita E, Muraki Y, et al. Reactivation of herpes simplex virus type 1 and varicella-zoster virus and therapeutic effects of combination therapy with prednisolone and valacyclovir in patients with Bell's palsy. The Laryngoscope. 2007;**117**:147-156

[69] Prichard MN, Shipman C Jr. A three-dimensional model to analyze drug-drug interaction. Antiviral Research. 1990;**14**:181-206

[70] Prichard MN, Aseltine KR, Shipman C Jr. MacSynergy™ II (Version 1.0). *Users Manual*. Ann Arbor: University of Michigan; 1992

[71] Vilhelmova-Ilieva N, Deffieux D, Quideau S, Galabov AS. Castalagin: Some aspects of mode of ant-herpes virus activity. Annals of Antivirals and Antiretrovirals. 2018;**2**(1):004-007

Section 2

Extraction

Chapter 4

Tannins: Extraction from Plants

Dang Xuan Cuong, Nguyen Xuan Hoan, Dinh Huu Dong, Le Thi Minh Thuy, Nguyen Van Thanh, Hoang Thai Ha, Dang Thi Thanh Tuyen and Dang Xuan Chinh

Abstract

The chapter presents mainly on different extraction methods of tannin. Some technical means required for effective extraction are also presented, for example, collection and treatment of plant and drying and storage of plant. Opportunity and challenges in application of extraction methods are also exhibited in the chapter.

Keywords: extraction, sea plant, tannin, terrestrial plant

1. Introduction

Tannins are high molecular weight phenolic compounds commonly found in plants with molecular weights ranging from 500 to over 3000 Da and up to 20,000 Da. Chemical structure of tannin is very diverse and different. More than 8000 different tannins have been detected. Tannins have been found in both free and bound forms in plant cells. Terrestrial plant tannins can be divided into four major groups: gallotannins, ellagitannins, proanthocyanidins (condensed tannins), and complex tannins. Sea plant tannins have been described as "phlorotannins," including oligomers or polymers of phloroglucinol. Tannin content can range from 0.2 to 25% DW [1], depending on plant species, harvest time, habitat of plants, and extraction method.

Many various bioactivities of tannins have also demonstrated including antioxidant, antibacterial [2], antifungal [3], antitumor, etc. The number of hydroxyl radicals and aromatic rings is important parameter determining in bioactivity of tannins. Ortho-dihydroxyl groups of tannin are important feature for chelate metal ions. High degree of polymerization and molecular weight play an important role for antioxidant activity of tannins. Tannins can bind tightly to protein through hydrogen bonding between the phenolic groups of the tannins and the -NH groups of peptides. These hydrogen bonds cannot be broken down by digestive enzymes or the attack of microorganisms [4].

Currently, tannins have been applied into many different fields including medicine, food, beverage, manufacture of ink and adhesives, dye and tanning industry, plastic resins, water purification, and surface coatings. Their applicability depends on its concentration, as a complexing agent or a precipitating agent [5].

From the above problems, it is necessary to extract and use tannins effectively. Therefore, this chapter focuses on extraction methods of tannins including maceration, microwave, ultrasound, enzyme, decoction, irradiated radiation, and gamma and points out some issues related to extraction efficiency of tannins, for example,

 IntechOpen

harvesting, handling, and storing materials. Opportunity and challenge of tannin extraction are also presented in the chapter.

2. Harvest and treatment of sample

2.1 Sea plants

Sea plants including seagrass, seaweeds, and mangroves contain a large amount of active tannins. The only way of seagrass collection is to dive into the sea. The harvest of seaweed has two ways: the first way, diving into the sea to collect seaweed; the second way, using machines for the harvest. Mangroves can only use an equipment for the harvest. Marine plant harvest should be selected to avoid the destruction of marine vegetation. Their physiological respiration ability is very strong, and they contain large amount of salts causing quick decay of sea plants immediately after the harvest. Hence, sea plants should be washed with salt water, then with fresh water moving salts. Sea plants are dried until 19% humidity by using freeze drying, infrared-freeze drying [6], microwave-vacuum drying, assisted microwave-vacuum drying [7], or hot air after cleaning seaweed, and infrared-freeze drying is usefully evaluated [8]. Infrared radiation power intensity of 5 $\frac{kW}{m^2}$ is suitable.

2.2 Terrestrial plants

Terrestrial plants are more diverse than sea plants, so the method of harvesting depends on the type of plant. The structure of terrestrial plants is more stable than marine plants, and they also do not contain salts. They still have a biochemical respiration. Therefore, they should also be dried by the above methods immediately after harvest. After drying, they also need to be grinded like marine plants for study and production. Terrestrial plants and marine plants should be stored in small plastic bags under refrigeration, the thing help longer storage time.

3. Role and position of tannins in plant cells

3.1 Sea plants

Tannins (phlorotannins) in sea plants exist in integral structural of cell walls. They directly participate in the structure of cells and bind to alginate, protein, laminarin, and fucoidan. Tannins also play an important role in the formation of the zygote's cell wall [9]. They have secondary ecological roles such as against UV radiation and grazing [10]. They possess metal sequestration capacity and algicidal effectiveness [11].

3.2 Terrestrial plants

Tannins have an important role in terrestrial plants, for instance, against microbial pathogens, harmful insects, and mammalian herbivores. They help in plant growth via binding to protein. They are involved in the cell structure of plants [12]. In all vacuoles and surface wax of plants, chloroplast-derived organelle and the ta nnosome produce tannins. They often exist in organelles where they are less affected by the protein precipitation. In Japanese persimmon fruits, the accumulation of tannin occurs in the vacuole of the tannin cells [12, 13], and tannin/tannin-less vacuoles are found in *Mimosa pudica*. Tannins are not accumulated in the vacuoles of sensitive plants [14].

DOI: http://dx.doi.org/10.5772/intechopen.86040

4. Extraction methods

4.1 Method of maceration

Maceration is one of the techniques used for tannin extraction from medicinal plants. Maceration is the simplest technique of extraction where the plant powder is placed in a closed vessel and soaked with the corresponding amount of solvent for a specified period of time until the tannins are dissolved in the solvent. In the first stage of the maceration, osmose occurs before diffusion occurs, in later stages, osmose occurs simultaneously with diffusion. The first stage is usually very short, and it can be a few seconds or hundredth second. The first-stage time depends on the type of solvent, the extraction material, and the extraction conditions. The movement of solvents in osmotic process from outside the cell into the cell and vice versa complies with van't Hoff's law. Dissolving solutes in osmotic process cause a pressure deficit called osmotic pressure. The osmosis is calculated on the basis of thermodynamics. Therefore, the osmotic pressure in ideal solutions is determined by Eq. (1), and the equation includes an osmotic coefficient (φs) for non-ideal solutions (Eq. 2). The osmotic coefficient depends on characteristic of each solute.

$$\pi = RT\sum Cs \quad (1)$$

$$\pi\text{observed} = RT\sum \varphi s Cs \quad (2)$$

where T is the temperature; R is the universal gas constant; and Cs is the molar concentration ($C = \frac{n}{V}$).

However, the actual pressure depends on the interaction between the solute and the cell membrane. Therefore, πobserved requires the reflection coefficient, σs:

$$\pi observed = RT\sum \sigma s \varphi s Cs \quad (3)$$

The osmose stops only when thermodynamic equilibrium occurs. It minimizes Gibbs free energy, $\Delta G = 0$, and temperature and pressure of solvent are fixed. Gibbs free energy is calculated by following equation:

$$G = E + \rho V - TS \quad (4)$$

where E is the energy; ρ is the pressure; V is the volume; S is the entropy ($S = k_B \ln \Omega$) inside the Boltzmann constant (kB); and number of configurations ($\ln \Omega$).

In osmotic process, osmotic pressure and hydrostatic pressure appear together playing an important role in fluid flow across the membrane. Characteristic parameters such as the permeability, the hydraulic conductivity, and the reflection coefficient reflect passive material transfer across cell membranes. Osmolarity is a concentration measure including freezing point depression, vapor pressure depression, and boiling point elevation. The diffusion in process of extraction depends on Fick's laws and Maxwell-Stefan diffusion. Therefore, the extraction yield and diffusion of tannin depend on material size, time and temperature of extraction, type of solvent, stirring, and size of tannin. Usually, the material size is optimal from 30 to 40 mesh [15]; the smaller material size creates difficulty during filtration. The material size depends on chemical characters of material and extraction method. In microwave-assisted method, material size larger than 50 meshes is suitable for the extraction and the filtration [16]. A modern mathematical exhibits Fick's first law and Fick's second law of diffusion as follow:

$$Fick's\, first \text{ law } N_i = -D_i \nabla c_i \tag{5}$$

$$Fick's\, second \text{ law } \frac{\partial c_i}{\partial t} = D_i \nabla^2 c_i \tag{6}$$

where N_i is the molar flux of i (mol $\frac{m^2}{s}$); D_i is the diffusion coefficient of i ($\frac{m^2}{s}$); and c_i is the concentration of i ($\frac{mol}{m^3}$).

Fick's second law exhibits the chemical species concentration being the dependent variable and diffusion of each chemical species occurring independently. When modeling diffusion of tannins, all diffusion coefficients are assumed equal and independent of temperature, pressure, etc. Fick's second law also exhibits a relation between the elapsed time and the square of diffuse distance. Fick's law is suitable for the solutions of one chemical species. Maxwell-Stefan diffusion describes good for multi-component diffusion. In Maxwell-Stefan diffusion, the species mole or mass fractions (xi and ωi, respectively) are dependent variable, and the mass flux of one chemical species is related to the concentration gradients of all chemical species. Parameterizing the diffusion rate of n components in a solution needs $n(n\text{-}1)/2$ independent coefficients. Total diffusion is equal to the total of molecular diffusion and convection term. Convection term, eddy diffusion, and molar flux due to convection are equal to each other [17, 18]. Convection term is calculated by the following equation:

$$\text{Convection term} = \text{concentration} \times \text{mass transfer velocity} = \text{CiV} \tag{7}$$

where Ci is the concentration of i ($\frac{\text{kmol}}{\text{m}^3}$) and V is the volume ($\text{m}^3$).

$$\text{Mass transfer velocity} = \frac{\text{mass flux}}{\text{concentration}} = \frac{\text{N}_\text{i} + \text{N}_\text{j}}{\text{C}_\text{T}} \frac{\frac{\text{kmol}}{\text{m}^2.\text{s}}}{\frac{\text{kmol}}{\text{m}^3}}; \frac{\text{m}}{\text{s}} \tag{8}$$

Thus, writing total diffusion equation for solution is

$$\text{Total diffusion} = \text{Ni} = \text{Ji} + \text{CiV} \tag{9}$$

$$\text{N}_\text{i} = -\text{D}_\text{ij}\frac{\text{dC}_\text{i}}{\text{dz}} + \frac{\text{C}_\text{i}}{\text{C}_\text{T}}(\text{N}_\text{i} + \text{N}_\text{j}) \tag{10}$$

where Ji is the molecular diffusion flux of i, ($\frac{\text{kmol}}{\text{m}^2.\text{s}}$); DAB is the diffusivity or diffusion coefficient for i in j, ($\frac{\text{m}^2}{\text{s}}$); and z is the distance of transfer (m).

The diffusion finishes only when mass balance of tannins between inside and outside the cell occurs. The stirring in the extraction process helps to increase the diffusion of tannins into the extract. When the mass balance of tannins occurs, new solvent is replaced to increase the efficiency of tannin extraction.

The technique is considered as a simply and low-cost way to extract tannins from medicinal plants. The technique is also a simple and popular choice for researchers because of non-complicated utensil and equipment. However, the duration of extraction time is long, the ratio of solvent to material is much, the extraction efficiency is not high, the volume of extraction tanks and the factory area are wide, compared to different techniques. The yield of tannin extraction is different when the techniques are different [19]. The yield of tannin extraction increases when hydrogen bonds between tannins and proteins are broken [20], or interactions between tannins and proteins during extraction are not formed [21, 22].

The yield of tannin extraction depends on many factors, for example, type of solvent (Snyder's solvent polarity index, pK_a, pK_b), plant species, temperature and time during the extracting process, and pressure in vessel. In the maceration, temperature of extraction can be hot, cold, or warm due to tannin characters and sample. In cold extraction, it requires thermolabile compounds. The hot extraction recommended for high temperature leading to decomposition proteins and cell structure during heating. When tannin extraction from *Galium tunetanum* Poiret, the experiment showed acetone being a better solvent, compared to ethanol (30%) [19], Na_2SO_3 solutions, NaOH solution, etc. [23]. When extracting tannins from Moroccan *Acacia mollissima* barks, methanol is evaluated being the best solvent at room temperature, and ethanol exhibits as the best solvent at 60°C [24]. Less tannin is extracted with aqueous or acidic methanol than with aqueous acetone [25]. For fresh barks of *Acacia mangium*, 50% acetone brings high yield of tannin extract [23]. The choice of the solvent types and the extraction temperature depends on the properties of the tannin in each plants. For extracting tannins from seaweed, ethanol is a best solvent because of high purified tannins and good activity. Temperature of extraction, type of solvent, and time of extraction are related to each other, and the relation exhibits clearly in **Table 1**.

Condensed tannins from grape pomace were extracted at 10°C for 120 min. The compound was then cooled, washed, and filtered. Adjusting the pH of the liquid

Solution	Condition of extraction	Reference
Ethanol	10 g of powdered plants in 100 mL of 30% ethanol at 60°C for 2 h.	[26]
Acetone	20 g of powdered plants in 200 mL of acetone for 24 h.	[27]
50% acetone	5 g of materials in 100 ml at room temperature for 24 h.	[23]
Water		
50% methanol	5 g of materials in 100 ml at room temperature for 24 h.	
Water	5 g of materials in 100 ml at 100°C for 40 min.	
Water	5 g of materials in 100 ml for 40 min, in autoclave	
2% Na_2SO_3		
4% Na_2SO_3		
6% Na_2SO_3		
0.04% NaOH solution	5 g of materials in 100 ml at 100°C for 40 min, in water-bath	
50% methanol	5 g of materials in 100 ml at 100°C for 4 h, reflux method	
50% methanol + Na_2SO_3 2%		
50% acetone		
50% acetone + Na_2SO_3 2%		
Water	Solvent and mangrove fruit ratio of 5:1 (w/w), 80°C for 60 min.	[28]
NaOH 5%	Solid-to-liquid ratio of 1:10 (w/v) at 10°C during 120 min	[29]
Water	10 g of sample in 200 ml at room temperature for 3 days	[30]
Ethanol		
50% (v/v) ethanol		

Table 1.
The conditions of tannin extraction in some literatures.

phase continuously, it reaches 1.5 by using dilute HCl for tannin precipitation. Finally, precipitated tannins were obtained at 8000 rpm for 15 minutes by centrifugation. Yield of extraction is higher than treatment of grape pomace by a solution of hydroxide sodium 5% (w/w) [29]. Maceration is similar to infusion method by which the materials are soaked in cold or boiling water for a short time [31].

4.2 Method of decoction

The decoction is similar to maceration method. However, the mixture is continuously heated at 100°C during decoction. Reaction kinetics due to the heat is stronger than the method of maceration. For extracting condensed tannins, the decoction is more effective than the maceration. In these methods, condensed tannin concentration depends on polarity of the extraction solvent. The mixture is then cooled to room temperature and the filtrate collection. Aqueous is a good solvent for the extraction of condensed tannins by decoction and maceration [32]. The decoction is very suitable for non-destroyed substances by the heat [31]. In addition to diffusion pressure and osmosis pressure, tannins are also separated from the material by the energy supplied from the thermal energy of the solvent and the affinity of the solvent. The thermal energy of the solvent separates tannins from the cells through the action on the bonds between tannins and cells. The affinity of the solvent will compete and pull tannins out of the material.

The method requires easy and cheap equipment; it is suitable for all production scales. However, the method consumes more energy than the maceration, and it is not suitable for the extraction of heat sensitive tannins [31].

4.3 Pressurized water extraction

The method is similar to infusion, hydrodistillation, decoction, and maceration. However, high pressure from 100 to 150 bar is used in the method, the thing is not found in different methods. Extraction temperature can be range from 60 to 100°C for 30 min static extraction. The solvent of water is commonly used in the method [33], and the other method includes subcritical water extraction, superheated water extraction, pressurized liquid extraction, and accelerated solvent extraction [34–36]. Pressurized water extraction is divided into static pressurized water extraction and dynamic pressurized water extraction. The residence time of tannins in dynamic pressurized water extraction is shorter than in static pressurized water extraction, and tannins are less degraded in dynamic pressurized water extraction. Hence, the extraction efficiency and the degradation of tannins depend on the extraction time [37]. The distribution ratio of tannins into the water plays an important role in the efficiency of tannin extraction. Dynamic pressurized water extraction is a higher investment, which is more difficult to use than static pressurized water extraction [38]. The temperature of extraction is often from the atmospheric boiling point of water (100°C/273 K, 0.1 MPa) to the critical point of water (374°C/647 K, 22.1 MPa) [38].

4.4 Ultrasound method

The method of ultrasound is ultrasonic-assisted maceration method commonly used for the extraction of tannins from plants, microalgae, and seaweeds, for instance, condensed tannins, hydrolysable tannins, and valonea tannins. In the method, ultrasonic power, temperature, and time of ultrasound are the factors that strongly influence the yield of tannin extraction. Obtained tannins by

Extraction method	The solvents				
	Acetone	96% ethanol	Ethyl acetate	Chloroform	n-Hexane
Ultrasound	1.618	2.112	0.277	0.222	0.058
Maceration	3.104	4.102	0.527	0.482	0.133
Soxhlet	2.213	3.715	0.283	0.261	0.095

Table 2.
Extracted tannin content from brown algae by the solvents and methods of difference.

ultrasonic-assisted maceration method are 17.6% higher than traditional methods [39]. In the method, the efficiency also depends on sound waves [40], which produce the mechanical vibrations in the solid causing implosion of cavitation bubbles and formed sheer forces. Thus, cell structure of plant is destroyed, and the yield of tannin extraction is highly increased. The acoustic waves are often most commonly measured more than >20 kHz [40, 41]. Ultrasound-assisted extraction is the technology of low investment and high efficiency for the extraction of tannins [42]. Ultrasound–assisted extraction method leads to efficient improvement of mass transfer and tannin extraction yield, reducing consumption of time and solvent, compared to the conventional method [43]. Ultrasound-assisted extraction method is environmentally friendly method, less risk of chemistry and physics, and less impact on molecular structural properties of tannins in plants [44, 45].

In the last decade, numerous publications on the use of ultrasound in active tannin extraction have been found. Ultrasonic frequency is commonly used being 40 kHz, ultrasonic time is about 30 min, and ultrasonic temperature can be a room temperature or warm temperature (55°C) [46, 47]. Analysis results of tannin content in brown algae commonly growing in Vietnam showed that the yield of tannin extraction by maceration is better than ultrasound and soxhlet. The thing is in contrast to tannins in terrestrial plants, and the expression of tannins in seaweed is more easily destroyed than tannins in terrestrial plants. Ninety-six percent ethanol is determined being the best solvent for tannin extraction from brown algae (**Table 2**), and the thing showing tannins in brown algae has a strong polar.

Some publications showed that ultrasonic frequency of 20 kHz and ultrasonic power ranging from 30 to 200 W are used for the extraction of tannins [39].

The dissolution rate of tannins from the plant material can be presented by Eq. (11) [48].

$$\frac{dC_t}{dt} = k(C_s - C_t)^2 \tag{11}$$

where Ct is the concentration of tannins in the extract at a time t (min), (mg $\frac{GAE}{L}$); Cs is the saturated concentration of tannins in the extracts, (mg $\frac{GAE}{L}$); and k is the second-order extraction rate constant, $\left(\frac{\frac{L}{g}}{\min}\right)$.

For determining kinetic parameters, some assumptions are given, t = 0 to t, Ct = 0 to Ct, the equation is written in the form of Eq. (12).

$$C_t = \frac{C_s^2.k.t}{1 + C_s.k.t} \tag{12}$$

where k is calculated according to the Arrhenius law

$$k = k_0 \exp\left(\frac{-E_a}{RT}\right) \tag{13}$$

where E_a is the value of the activation energy; R is the gas constant ($8.314 \frac{J}{mol}$ K); and T is the temperature of the extraction (K).

Ultrasonic-assisted extraction is also combined with ionic liquid improving extraction efficiency [49].

4.5 Microwave-assisted extraction

The microwave-assisted extraction is a good choice for the extraction of tannins using microwave (electromagnetic radiations in the frequency range from 300 MHz to 300 GHz) energy [50]. The heating mechanism of microwave is very special [51]. Ionic conduction and dipole rotation cause the transformation of the microwave energy to heat through interacting with polar molecules [50]. The microwave supplies the high temperature (100–150°C) in the extraction process [52]. The yield of tannin extraction strongly depends on the penetration depth of microwaves into plants and the power of microwaves. Inside the dielectric constant, the moisture content and temperature of plants, and the frequency of the electrical field decide the penetration of microwave depth. The water in plants absorbs microwave energy to reach the superheated state and cell structure disruption. The thing is to help to increase the diffusion of tannins into the extract. The solvents such as water, ethanol, and methanol strongly absorb microwaves energy, and their dielectric constant is high, compared to toluene or hexane. Toluene and hexane maintain the surrounding extraction solvent in the cold state because of their dielectric constant. Water, ethanol, and methanol are advised for the extraction of tannins, because tannins are very soluble in them [53]. Microwave-assisted maceration is a simple, time-saving method, less cost of solvent, and energy. The microwave technique brings the higher efficiency of tannin extraction than ultrasound method [54, 55]. The yield of tannin extraction in microwave method is 1.25 times more than maceration method [24]. The efficiency of tannin extraction is arranged in descending order according to the following methods: microwave, maceration, and infusion. Microwave-assisted extraction is less efficiency for non-polar tannins and non-polar solvents [56].

4.6 Ionic liquid-based microwave-assisted extraction

Ionic liquid-based microwave-assisted extraction is a high efficient combination method in tannin extraction from plants. Ionic liquid aqueous solution-to-material ratio of 20:1 is suitable in the method. Currently, there are numerous ionic liquids, for example, [Emim]Br, [Bmim]Br, [Hmim]Br, [Omim]Br, [Dmim]Br, [Bmim]Cl, [Bmim]BF4, [Bmim]NO3, and [Bmim]OH, where Emim is 1-ethyl-3-methylimidazolium; Bmim is 1-butyl-3-methylimidazolium; Hmim is 1-hexyl-3-methylimidazolium; Omim is 1-octyl-3-methylimidazolium; and Dmim is 1-decyl-3-methylimidazolium.

In the method, the materials are soaked in ionic liquid for 3 h before microwave-assisted extraction. Extraction time of tannins by microwave is about 10 min under microwave power of 230 W. A 1.25 M sodium chloride and 80% ethanol are used as the solvent in ionic liquid-based microwave-assisted extraction [57]. This is a friendly method and low energy consumption but high cost.

4.7 Infrared-assisted extraction

Infrared-assisted extraction is a maceration using an infrared lamp to supply the heat, and the penetration of infrared into the material is higher than other method [58]. In conventional methods, the solvent is heated before heating the material. In infrared method, the material is directly heated without heating the solvent [59]. The effectiveness of the infrared method depends on the absorption characteristics of the extracting solvent, the wavelength of the infrared heater, and the distance between the lamp and the material. Electromagnetic waves excite tannins of material in the modes of twisting, stretching, and bending, increasing the efficiency of tannin extraction [59, 60]. Infrared method exhibits a higher extraction yield than microwave-assisted extraction and ultrasound extraction [41]. Infrared-assisted extraction is low cost, but extraction efficiency of tannins from plants is high, and the time of infrared extraction is about 30 min [60].

4.8 Soxhlet extraction

Extraction method and solvent play an important role in the extraction of tannins. Soxhlet extraction is a good method when compared to cold maceration. The effectiveness of soxhlet extraction is mainly based on the evaporation temperature and polarity index of the solvent, and some characters of solvents are presented in **Table 3**. The suitable temperature of extraction improves the diffusion ratio of tannins into the solvent and the circulation of the solvent. Fifty percent ethanol gives the highest extract yield of tannins from herbal plants [30]. Soxhlet extraction with water content is also evaluated as the best method for gallic and ellagic acid extractions. Solvent polarity, solvent-to-solid ratio, and contact time significantly affect the yield of tannin extraction [33]. The solvents such as water or ethanol-water mixtures are used for the extraction of the active hydrolysable tannins (gallic acid and ellagic acid). Extract yield of tannins corresponded to 27.1, 26.4, 26.2, 22.5, 14.6, and 11.6 (% g/g sample) when using extraction solvent of ethanol:water (20%:80%), ethanol:water (30%:70%), water, ethanol:water (50%:50%), methanol, and ethanol, respectively, in the method [33]. Extract yield of tannins is the lowest with n-hexane. Synder's solvent polarity index almost causes high extract yield of tannins. Synder's solvent polarity index is calculated according to the following equation [61]:

$$\text{Synder's solvent polarity index} = \left(\frac{I_A}{100}.P_A\right) + \left(\frac{I_B}{100}.P_B\right) \tag{14}$$

Extraction solvent	Synder's solvent polarity index	Boiling point (°C)	Extraction solvent	Synder's solvent polarity index	Boiling point (°C)
n-Hexane	0.1	69	70% acetone, 30% water	6.5	84
Dichloro-methane	3.4	40	70% ethanol, 30% water	8.2	90
Chloroform	4.1	61	50% ethanol, 50% water	7.9	94
Acetone	5.4	56	30% ethanol, 70% water	7.1	97
Ethanol	5.2	78	20% ethanol, 80% water	6.3	98
Methanol	6.6	65	Water	9.0	100

Table 3.
Synder's solvent polarity index and boiling point of some solvents.

where I_A and I_B are the polarity index of solvents A and B, respectively; P_A and P_B are the percentage of solvents A and B, respectively.

However, soxhlet method needs long duration and high amount of solvent, hence the loss efficiency of economic and the environmental problems. When the evaporation temperature of the solvents is high, thermal destruction of tannins can happen [53]. Extraction time is about 3 h or 6 h due to the materials [33].

4.9 Supercritical fluid extraction

The supercritical fluid extraction is a modern and environment friendly method, which maintains the chemical and biological characters of tannins. The supercritical solvents have an important role in the efficiency of tannin extraction, and **Table 4** presents some properties of useful solvents for tannin extraction. The supercritical solvents are used as extraction solvent in the method including carbon dioxide, hexane, pentane, butane, nitrous oxide, sulfur hexafluoride, and fluorinated hydrocarbons. Supercritical carbon dioxide is often the choice because of its low critical pressure (100 and 450 bar) [62]. However, the selectivity of CO_2 is not good. For improving the problems, a co-solvent or modifier is combined with CO_2 in tannin extraction [63]. Water and ethanol-water cosolvents exhibit a good result in the separation of the less polar compounds and the hydrolysable tannins [53]. The diffusion coefficient of the supercritical fluid is higher than a liquid solvent, but the viscosity and the surface tension of the supercritical fluid are lower than a liquid solvent. Hence, the diffusion and mass transfer of tannins are better when using the supercritical fluid. The solvation power of supercritical fluid can be changed via the change of temperature and/or pressure [53, 56].

From the parameters in **Table 4** and previous claims, it is easy to see that the higher the viscosity of the solvent is, the greater the solubility of tannins is.

4.10 Enzyme-assisted extraction

Enzyme-assisted extraction is one of the environmentally friendly methods using an enzyme or complex enzymes to improve bioactive extraction yield via the disruption of the material cell. This method is combined with various methods, but high temperature is not used for the extraction of tannins because protein is precipitated under high temperature. Extraction time or time of enzyme treatment is sufficient and not long because a long time leads to release tannins and the formation of protein-tannin complexes. Of course, not all tannins precipitate all proteins, this precipitation is selective. The tannin-protein complexes are only formed when

Solvent	The moment of dipole (Debye)	The constant of the dielectric	Cohesive energy density ($J\ \frac{mol}{mL}$)	Viscosity (mPa)	The tension of surface ($\frac{Cal}{mol}$)
n-Hexane	0.00	1.88	200.76	0.30	25.75
Acetone	2.88	20.49	362.07	0.31	33.77
Chloroform	1.04	4.71	332.00	0.54	38.39
Ethanol	1.69	24.85	618.87	1.07	31.62
Methanol	1.70	32.61	808.26	0.54	31.77
Water	1.87	78.36	2095.93	0.89	104.70

Table 4.
Some properties of solvents used for tannin extraction [64].

proteins and tannins have high molecular weight, an open flexible structure, hydrophobic proteins of proline richness [65, 66]. The linkages between tannins and proteins are unstable. The linkages between proteins and tannins are formed from phenolic groups of tannins and the carbonyl groups of peptides [65]. Four types of bonds between proteins and tannins are suggested in **Table 5** [67].

For enhancing the extraction effectiveness, the treatment of the material cell by the enzyme complex should be used prior to tannin extraction. Enzymatic pretreatment achieves a reduction in the extraction time and energy consumption [68]. Enzymes complexes weaken or break down the linkages between tannins and cell wall [69–71]. In this way, the partial or overall degradation of material cells allows leasing tannins from plants [72, 73]. The cell-wall degrading enzymes can be used for tannin extraction including Celluclast® 1.5 L, Pectinex® Ultra, Novoferm®, and hemicellulases. The treatment of seaweed by three different enzymes shows that the efficiency of the cell membrane of seaweed is significantly different. Viscozyme L enzyme destroys the seaweed cells better than termamyl enzyme (**Figures 1** and **2**), and eroding ability of seaweed cell membrane of cellulase enzyme at 60°C is better at 50°C (**Figures 3** and **4**).

Cellulase enzyme usually exhibits high efficiency in tannin extraction from seaweed at other temperatures, compared to other enzymes (**Figure 5**). Enzyme-assisted extraction permits the extraction of tannins with high stability and high activity. This is fully feasible because catalyze reactions and regioselectivity of enzymes can take place in aqueous medium under mild conditions [74]. Disrupting the structural integrity of the material cells by enzymes is the easy method and low solvent consumption. However, a clear understanding of the catalytic property, the material structure, the mode of action, and the enzyme conditions is

Types of bonds	Interactions	Control factor
Hydrogen bonds	The oxygen of amide groups in the peptide bonds of proteins and the hydroxyl radicals of the tannins	Reversible and dependent on pH
Hydrophobic interactions	The hydrophobic regions of the proteins and the aromatic rings of the tannins	Reversible and dependent of pH
Ionic bonds	The cationic sites of the proteins and the phenolate anion	
Covalent bonding	The oxidation of polyphenols to quinones and their subsequent condensation	Irreversible

Table 5.
Some types of bonds between proteins and tannins.

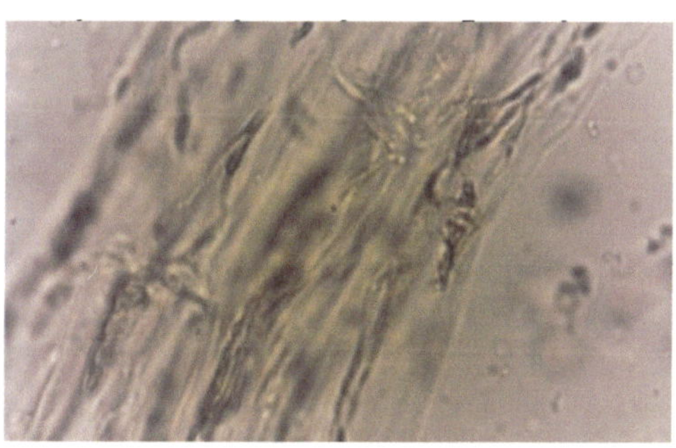

Figure 1.
A part of algae stem Sargassum olygocystum after being treated at 40°C with 0.3% enzyme Termamyl.

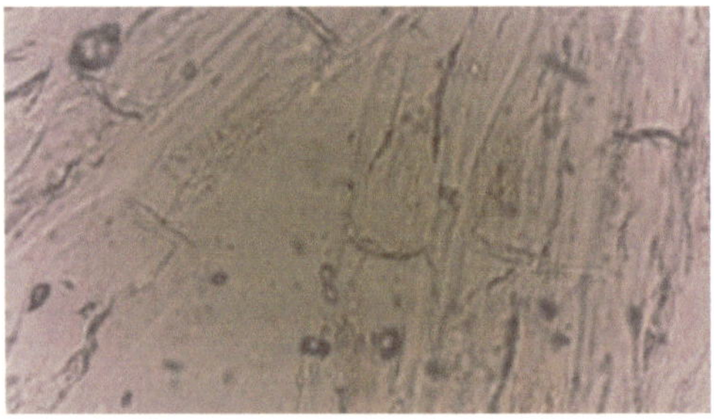

Figure 2.
A part of algae stem Sargassum olygocystum after being treated at 40°C with 0.3% enzyme Viscozyme L.

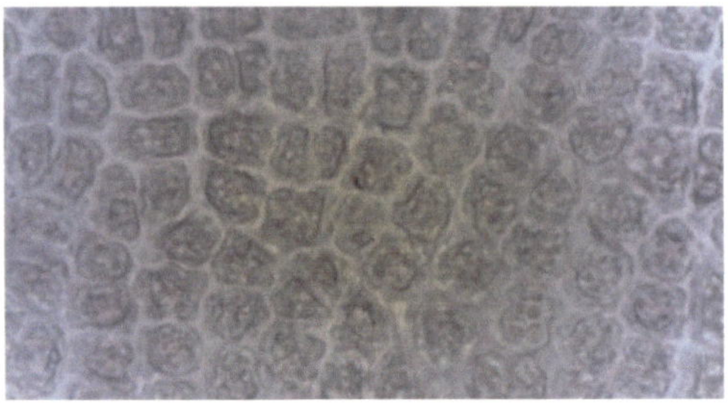

Figure 3.
A part of algae leave Sargassum olygocystum after being treated at 50°C with 0.3% enzyme cellulase.

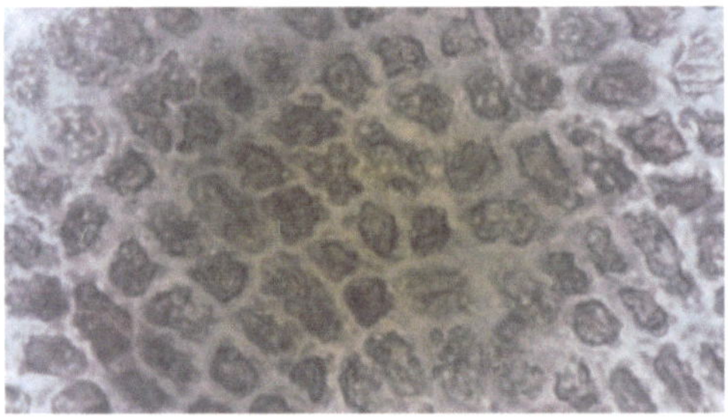

Figure 4.
A part of algae leave Sargassum olygocystum after being treated at 60°C with 0.3% enzyme cellulase.

necessary [68]. The effectiveness of tannin extraction mainly depends on ionic strength [75], ions in solution [76], and characters of proteins and tannins. Some problems on technology need to consider before large production, for example, the price of enzymes, hydrolyze capacity, and operating conditions of enzymes [68].

4.11 Gamma-assisted extraction

Gamma-assisted extraction is a new method using Co-60 gamma radiation for tannin extraction. Co-60 gamma radiation increases the effectiveness of the tannins from *Anacardium occidentale* [77]. However, this method still needs more study to apply, and expanding ability in large scale also entails many legal issues. For tannin extraction, materials can be directly irradiated by gamma radiation, materials are then soaked in specific solvent, or the mixture of materials and solvents is directly irradiated by gamma radiation. Radiation dose of 5–25 kGy is concordant with tannin extraction from brown algae.

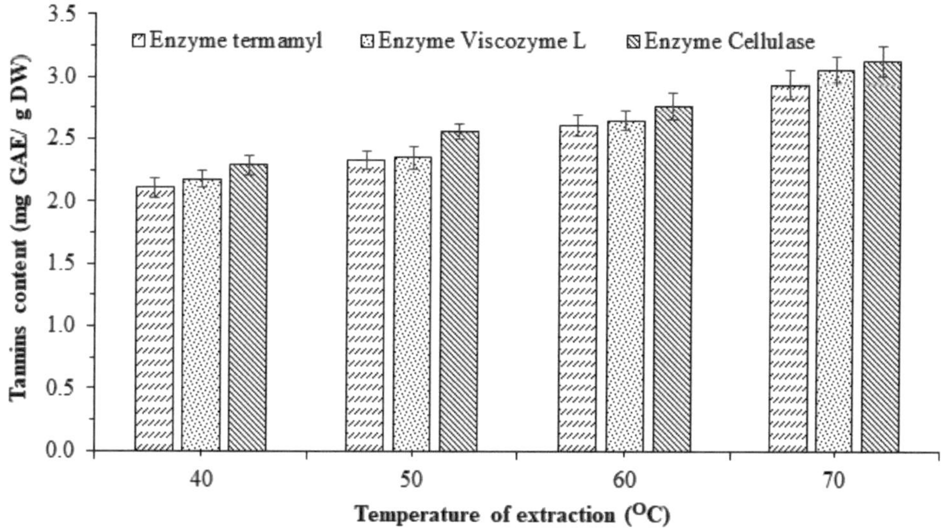

Figure 5.
Extracted tannin content from brown algae in different conditions of enzyme and temperature.

5. Opportunity and challenge of tannin extraction

Tannins commonly exist in the nature, and they are bioactive diversity such as antioxidant, antifungal, anti-virus, anti-inflammatory, and anti-parasitic. Tannins are used in numerous fields, for example, feed of ruminants, monogastric animals, etc. Each method of tannin extraction has difficult and easy points, for example, high investment, difficult operation, or popular equipment. However, all methods are based on the dynamic and static maceration. All modern techniques are supporting and improving traditional extraction methods. At the same time, the demand using tannins in the market is very huge. Therefore, opportunity of tannin extraction and commercializing tannins are great.

However, large-scale production of tannins greatly differs in comparison to small-scale production. The content, structure, and activity of tannins in plants depend on growth time of plants, habitat place of plants, species, the parts of the plants, etc. [78]. Hence, stabilizing the content, structure, and activity of tannin is difficult. Tannin content is decreased in storage time of materials. The results of the tannin activities are mainly at the laboratory level. At the same time, modern method of tannin extraction needs high investment, good manpower, and great market. There are some challenges in tannin extraction and its application into the life.

6. Conclusions

The chapter presents the position and the role of tannins in plant cell structure, the harvest and treatment of plant materials, the opportunities, and the challenges of tannin extraction from plants to apply into various fields. Specially, traditional and modern methods of tannin extraction are also discussed specifically including the extraction techniques, the advantages and the disadvantages of the method, the mechanism of the extraction process, the solutions to efficiency improvement of tannin extraction, some experimental parameters of tannin extraction, some conversion process in the extraction, mathematical model of transformation, and kinetics of extraction process.

The presented extraction methods have the advantages and the disadvantages, but they can be used effectively in tannin extraction from plants. The methods of tannin extraction should continue to be improved to bring higher efficiency.

Conflict of interest

Non conflict of interest declaration.

Abbreviations

DW	dry weight
GAE	gallic acid equivalent
g	gram
kGy	kilogray
Min	minutes
v	volume
w	weight
W	watt
M	mol

Author details

Dang Xuan Cuong[1*], Nguyen Xuan Hoan[2], Dinh Huu Dong[2], Le Thi Minh Thuy[3], Nguyen Van Thanh[4], Hoang Thai Ha[5], Dang Thi Thanh Tuyen[6] and Dang Xuan Chinh[7]

1 Nhatrang Institute of Technology Application and Research, VAST, Nhatrang, Khanhhoa, Vietnam

2 Ho Chi Minh City University of Food Industry, Ho Chi Minh City, Vietnam

3 College of Aquaculture and Fisheries, Can Tho University, Ninhkieu, Cantho, Vietnam

4 Kien Giang University, Chauthanh, Kiengiang, Vietnam

5 Ho Chi Minh City Industry and Trade College, Ho Chi Minh City, Vietnam

6 Faculty of Food Technology, Nha Trang University, Nhatrang, Khanhhoa, Vietnam

7 Pacific Ocean University, Nhatrang, Khanhhoa, Vietnam

*Address all correspondence to: cuong_mails@yahoo.com.vn

References

[1] Margarete D, Valerie M, Nadir F, Nadir E, James FC. Root condensed tannins vary over time, but are unrelated to leaf tannins. AoB PLANTS. 2018;**10**(4):ply044. DOI: 10.1093/aobpla/ply044

[2] Okuda T, Yoshida T, Hatano T, Ito H. Ellagitannins renewed the concept of tannins. In: Quideau S, editor. Chemistry and Biology of Ellagitannins: An Underestimated Class of Bioactive Plant Polyphenols. Singapore: World Scientific Publishing Co; 2009. pp. 1-54

[3] Morey AT, de SFC, Santos JP, Pereira CA, Cardoso JD, de ARS, et al. Antifungal activity of condensed tannins from stryphnodendron adstringens: Effect on *Candida tropicalis* growth and adhesion properties. Current Pharmaceutical Biotechnology. 2016;**17**(4):365-375. DOI: 10.2174/1389201017666151223123712

[4] Hagerman AE. Chemistry of tannin-protein complexation. In: Richard WH, Joseph JK, Susan JB, editors. Chemistry and Significance of Condensed Tannins. United States: Springer; 1989. pp. 323-333

[5] Antonio P. Chapter 8—Tannins: Major sources, properties and applications. In: Mohamed B, Alessandro G, editors. Monomers, Polymers and Composites from Renewable Resources. Elsevier Science, Elsevier Ltd; 2008. pp. 179-199

[6] Antal T. Comparative study of three drying methods: Freeze, hot airassisted freeze and infrared-assisted freeze modes. Agronomy Research. 2015;**13**(4): 863-878

[7] Rattapon S, Ampawan T, Gauri SM. Effect of far-infrared radiation assisted microwave-vacuum drying on drying characteristics and quality of red chilli. Journal of Food Science and Technology. 2015;**52**(5):2610-2621. DOI: 10.1007/s13197-014-1352-4

[8] Navin KR. Recent Trends and developments in infrared heating in food processing. Critical Reviews in Food Science and Nutrition. 2012;**52**(9): 737-760. DOI: 10.1080/10408398.2010.508138

[9] McInnes AG, Ragan MA, Smithand JAWDG. High-molecular-weight phloroglucinol-based tannins from brown algae: Structural variants. Hydrobiologia. 1984;**116-117**(1): 597-602. DOI: 10.1007/BF00027755

[10] Arnold TM, Targett NM. To grow and defend: Lack of tradeoffs for brown algal phlorotannins. Oikos. 2003;**100**(2): 406-408. DOI: 10.1034/j.1600-0706.2003.11680.x

[11] Nagayama K, Shibata T, Fujimoto K, Honjo T, Nakamura T. Algicidal effect of phlorotannins from the brown alga *Ecklonia kurome* on red tide microalgae. Aquaculture. 2003;**218**(1-4):601. DOI: 10.1016/S0044-8486(02)00255-7

[12] Claudia MF, Lucimar BM, Deborah YACS. Chapter 10. Tannins: What do they represent in plant life? In: Georgios KP, editor. Tannins: Types, Foods Containing, and Nutrition. New York: Nova Science Publishers Inc; 2010. p. 13

[13] Shinya K, Keizo Y, Akira S. Identification of molecular markers linked to the trait of natural astringency loss of Japanese persimmon (*Diospyros kaki*) fruit. Journal of the American Society for Horticultural Science. 2001; **126**(1):51-55

[14] Fleurat-Lessard P, Béré E, Lallemand M, Dédaldéchamp F, Roblin G. Co-occurrence of tannin and tannin-less vacuoles in sensitive plants. Protoplasma. 2016;**253**(3):821-834. DOI: 10.1007/s00709-015-0844-z

[15] Zhe-Xiong J, Bao-Qing W, Zhi-Jie C. Microwave-assisted extraction of tannins from Chinese herb *Agrimonia pilosa* Ledeb. Journal of Medicinal Plants Research. 2008;**4**(21):2229-2234. DOI: 10.5897/JMPR10.378

[16] Meghal D, Jigisha P, Parikh PA. Extraction of natural products using microwaves as a heat source. Separation & Purification Reviews. 2010;**39**(1-2): 1-32. DOI: 10.1080/15422111003662320

[17] Coulson JM, Richardson JF. Chemical Engineering. 5th ed. ELBS: Pergamon Press; 2002. 1209 p

[18] Chhabra RP. In: Backhurst JR, Harker JH, Richardson JF, Coulson JM, editors. Chemical Engineering. 6th ed. India: Butterworth-Heinemann; 1999. 928 p

[19] Sofiane G, Daoud H, Seoussen K, Saliha D. The comparison of two tannin extraction methods from *Galium tunetanum* Poiret and their antioxidant capacities. Der Pharmacia Lettre. 2014; **6**(1):114-119

[20] Porter LJ. Condensed tannins. In: Rowe JW, editor. Natural Products of Woody Plants. Berlin: Springer-Verlag; 1989. pp. 651-690

[21] Hagerman AE. The Tannin Handbook. Oxford: Miami University; 2002. 116 p

[22] Rizwan AS, Jeevan MBD, Venkat RN, Petricia RI, Sachin M, Ashish T. Hepatoprotective activity of alcoholic and aqueous extracts of bark of *Bassia Latifolia Roxb*. against paracetamol induce hepatotoxicity in rats. Der Pharmacia Lettre. 2012;**4**(4):1272-1284

[23] Wina E, IWR S, Tangendjaja B. Biological activity of tannins from *Acacia mangium* bark extracted by different solvents. Media Peternakan. 2010;**33**(2):103-107

[24] Rhazi N, Oumam M, Hannache H, Sesbou A, Charrier B, Pizzi A, et al. Comparison of the impact of different extraction methods on polyphenols yields and tannins extracted from Moroccan *Acacia mollissima* barks. Industrial Crops and Products. 2015;**70**: 245-252. DOI: 10.1016/j.indcrop.2015.03.016

[25] Hagerman AE. Extraction of tannin from fresh and preserved leaves. Journal of Chemical Ecology. 1988;**14**(2): 453-461. DOI: 10.1007/BF01013897

[26] Barrau E, Fabre N, Fouraste I, Hoste H. Effect of bioactive compounds from sainfoin (*Onobrychis viciifolia*) on the in vitro larval migration of *Haemonchus contortus:* Role of tannins and flavonol glycosides. Parasitology. 2005;**131**: 531-538. DOI: 10.1017/S0031182005008024

[27] Mohamad IMN, NorNadiah MY, Amirue AA. Extraction of tannin from oil palm empty fruit bunch as a rust deactivator. In: Regional Symposium On Chemical Engineering 2005; November 30th-December 2, 2005; Horison Hotel. Hanoi: Vietnam. 2005. pp. 197-201

[28] SubandriyoNanik IS. Extraction process for reducing tannin of mangrove fruit [*Bruguiera gumnorrhiza* (Lamarck, 1798)] as a raw material for food flour. Aquatic Procedia. 2016;7: 231-235. DOI: 10.1016/j.aqpro.2016.07.032

[29] Noureddine B, Mohamed A, Mohamed KS, Ramzi K. Properties of tannin-glyoxal resins prepared from lyophilized and condensed tannin. Journal of Textile Engineering & Fashion Technology. 2017;**3**(4):705-711. DOI: 10.15406/jteft.2017.03.00110

[30] Kamarudin NA, Markom M, Latip J. Effects of solvents and extraction methods on herbal plants *Phyllanthus niruri, Orthosiphon stamineus* and *Labisia pumila*. Indian Journal of Science and

Technology. 2016;**9**(21). DOI: 10.17485/ijst/2016/v9i21/95235

[31] Bimakr M. Comparison of different extraction methods for the extraction of major bioactive flavonoid compounds from spearmint (*Mentha spicata* L.) leaves. Food and Bioproducts Processing. 2010;**89**(1):67-72. DOI: 10.1016/j.fbp.2010.03.002

[32] Abbassia B, Hassiba MB, Imene S. Effects of different extraction methods and solvents on the phenolic composition and antioxidant activity of *Silybum marianum* leaves extracts. International Journal of Medical Science and Clinical Invention. 2018;**5**(3): 3641-3647. DOI: 10.18535/ijmsci/v5i3.16

[33] Masturah M, Masitah H, Wan RWD, Harcharan S, Jamaliah MJ. Extraction of hydrolysable tannins from *Phyllanthus niruri* Linn.: Effects of solvents and extraction methods. Separation and Purification Technology. 2007;**52**(3):487-496. DOI: 10.1016/j.seppur.2006.06.003

[34] Teo CC, Tan SN, Yong JWH, Hew CS, Ong ES. Pressurized hot water extraction (PHWE). Journal of Chromatography A. 2010;**1217**(16): 2484-2494. DOI: 10.1016/j.chroma.2009.12.050

[35] Miguel H, María CP, José AM, Elena I. Compressed fluids for the extraction of bioactive compounds. Trends in Analytical Chemistry. 2013;**43**:67-83. DOI: 10.1016/j.trac.2012.12.008

[36] Saldaña MDA, Valdivieso-Ramírez CS. Pressurized fluid systems: Phytochemical production from biomass. Journal of Supercritical Fluids. 2015;**96**:228-244. DOI: 10.1016/j.supflu.2014.09.037

[37] Liu J, Sandahl M, Sjöberg PJR, Turner C. Pressurised hot water extraction in continuous flow mode for thermolabile compounds: Extraction of polyphenols in red onions. Analytical and Bioanalytical Chemistry. 2014; **406**(2):441-445. DOI: 10.1007/s00216-013-7370-7

[38] Merichel P, Charlotta T. Pressurized hot water extraction of bioactives. Trends in Analytical Chemistry. 2015;**71**: 39-54. DOI: 10.1016/j.trac.2015.02.022

[39] Cigdem KO, Hasan O. Ultrasound extraction of valonea tannin and its effects on extraction yield. Journal of American Leather Chemists Association. 2012;**107**(11):394-403

[40] Mason TJ. Ultrasound in synthetic organic chemistry. Chemical Society Reviews. 1997;**26**:443-451. DOI: 10.1039/CS9972600443

[41] Dina C, Nada ED, Hiba NR, Iman EG, Youssef M, Adla J, et al. Study of the selectivity and bioactivity of polyphenols using infrared assisted extraction from apricot pomace compared to conventional methods. Antioxidants (Basel). 2018;**7**(12): 174-185. DOI: 10.3390/antiox7120174

[42] Sahin S, Aybastıer O, Işık E. Optimisation of ultrasonicassisted extraction of antioxidant compounds from *Artemisia absinthium* using response surface methodology. Food Chemistry. 2013;**141**(2):1361-1368. DOI: 10.1016/j.foodchem.2013.04.003

[43] Vilkhu K, Mawson R, Simons L, Bates D. Applications and opportunities for ultrasound assisted extraction in the food industry—A review. Innovative Food Science and Emerging Technologies. 2008;**9**(2):161-169. DOI: 10.1016/j.ifset.2007.04.014

[44] Chemat F, Zill EH, Khan MK. Applications of ultrasound in food technology: Processing, preservation and extraction. Ultrasonics Sonochemistry. 2011;**18**(4):813-835. DOI: 10.1016/j.ultsonch.2010.11.023

[45] Yuan J, Huang J, Wu G, Tong J, Xie G, Duan JA, et al. Multiple responses optimization of ultrasonic-assisted extraction by response surface methodology (RSM) for rapid analysis of bioactive compounds in the flower head of *Chrysanthemum morifolium*. Industrial Crops and Products. 2015;**74**:192-199. DOI: 10.1016/j.indcrop. 2015.04.057

[46] Jordana NS, Nathália BP, Leonardo LB, Gerlon ARO, José RP, Edemilson CC. Optimization of ultrasound-assisted extraction of polyphenols, tannins and epigallocatechingallate from barks of *Stryphnodendron adstringens* (Mart.) Coville bark extracts. Pharmacognosy Magazine. 2014;**10**(Suppl 2):S318-SS23. DOI: 0.4103/0973-1296.133287

[47] Miaomiao L, Jing W, Kun Y, Yiman Q, Jie Z, Mingtao F, et al. Optimization of ultrasonic-assisted extraction of antioxidant tannin from young astringent persimmon (*Diospyros kaki* L.) using response surface methodology. Journal of Food Processing and Preservation. 2018;**42**(7):e13657. DOI: 10.1111/jfpp.13657

[48] Lazar L, Talmaciu AI, Volf I, Popa VI. Kinetic modeling of the ultrasound-assisted extraction of polyphenols from *Picea abies* bark. Ultrasonics Sonochemistry. 2016;**32**:191-197. DOI: 10.1016/j.ultsonch.2016.03.009

[49] Yang L, Liu Y, Zu YG, Zhao CJ, Zhang L, Chen XQ, et al. Optimize the process of ionic liquid-based ultrasonic-assisted extraction of aesculin and aesculetin from Cortex Fraxini by response surface methodology. Chemical Engineering Journal. 2011; **175**(1):539-547. DOI: 10.1016/j. cej.2011.09.110

[50] Wang Y, Xi GS, Zheng YC, Miao FS. Microwave-assisted extraction of flavonoids from Chinese herb *Radix Puerariae* (Ge Gen). Journal of. Medicinal Plants Research. 2010;**4**(4): 304-308. DOI: 10.5897/JMPR

[51] Ming YH, Swee NT, Jean WHY, Eng SO. Emerging green technologies for the chemical standardization of botanicals and herbal preparations. Trends in Analytical Chemistry. 2013;**50**:1-10. DOI: 10.1016/j.trac.2013.03.012

[52] Tricia SR, Susan VO. Bridging the gap between supercritical fluid extractions and liquid ext raction techniques: Alternative approaches of the extraction of solid and liquid environmental matrices. Critical Reviews in Analytical Chemistry. 1996; **26**(2&3):1-39. DOI: 10.1080/10408349608050568

[53] Marie INN, Ebrahimi M, Dieudonne ML, Zacharie NDE, Doriane N. Review on extraction and isolation of plant secondary metabolites. In: 7th Int'l Conference on Agricultural, Chemical, Biological and Environmental Sciences (ACBES-2017); May 22–24, 2017. Kuala Lumpur: Malaysia; 2012. pp. 67-72

[54] Gallo M, Ferracane R, Graziani G, Ritieni A, Fogliano V. Microwave assisted extraction of phenolic compounds from four different spices. Molecules. 2010;**15**(9):6365-6374. DOI: 10.3390/molecules15096365

[55] Han ZP, Liu RL, Cui HY, Zhang ZQ. Microwave-assisted extraction and LC/MS analysis of phenolic antioxidants in sweet apricot (*Prunus armeniaca* L.) kernel skins. Journal of Liquid Chromatography & Related Technologies. 2013;**36**:2182-2195. DOI: 10.1080/10826076.2012.717057

[56] Wang L, Waller C. Recent advances in extraction of nutraceuticals from plants. Trends in Food Science & Technology. 2006;**17**(6):300-312. DOI: 10.1016/j.tifs.2005.12.004

[57] Lei Y, Xiaowei S, Fengjian Y, Chunjian Z, Lin Z, Yuangang Z. Application of ionic liquids in the microwave-assisted extraction of proanthocyanidins from *Larix gmelini*

bark. International Journal of Molecular Sciences. 2012;**13**(4):5163-5178. DOI: 10.3390/ijms13045163

[58] René E, René M, Joel M. Infrared irradiation: Toward green chemistry, a review. International Journal of Molecular Sciences. 2016;**17**(4):453-478. DOI: 10.3390/ijms17040453

[59] Yi C, Yingjia Y, Yingjia Y, Gengli D, Yan L. Study on infrared-assisted extraction coupled with high performance liquid chromatography (HPLC) for determination of catechin, epicatechin, and procyanidin B2 in grape seeds. Food Chemistry. 2011; **127**(4):1872-1877. DOI: 10.1016/j.foodchem.2011.02.026

[60] Chen Y, Duan G, Xie M, Chen B, Li Y. Infrared-assisted extraction coupled with high-performance liquid chromatography for simultaneous determination of eight active compounds in *Radix Salviae miltiorrhizae*. Journal of Separation Science. 2010;**33**(17-18):2888-2897. DOI: 10.1002/jssc.201000234

[61] Alan RK, Dan CF, Hongfang Y, Kaido T. Quantitative measures of solvent polarity. Chemical Reviews. 2004;**104**(1):175-198. DOI: 10.1021/cr020750m

[62] Yepez B, Espinosa M, López S, Bolaños G. Producing antioxidant fractions from herbaceous matrices by supercritical fluid extraction. Fluid Phase Equilibria. 2002;**194-197**:879-884. DOI: 10.1016/S0378-3812(01)00707-5

[63] Ankit G, Madhu N, Vijay K. Modern extraction methods for preparation of bioactive plant extracts. International Journal of Applied and Natural Sciences. 2012;**1**(1):8-26

[64] Chong-Hui G, Li H, Rajesh BG, Krishnaswamy R. Grouping solvents by statistical analysis of solvent property parameters: Implication to polymorph screening. International Journal of Pharmaceutics. 2004;**283**(1-2):117-125. DOI: 10.1016/j.ijpharm.2004.06.021

[65] Hagerman AE, Buttler LG. The specificity of proanthocyanidin-protein interactions. Journal of Biological Chemistry. 1981;**256**(9):4494-4497

[66] Mueller-Harvey I. Tannins: Their nature and biological significance. In: Caygill JC, Mueller-Harvey I, editors. Secondary Plants Products: Antinutritional and Beneficial Actions in Animal Feeding. United Kingdom: Nottingham University Press; 1999. pp. 17-70

[67] Ravindra K, Manohar S. Tannins: Their adverse role in ruminant nutrition. Journal of Agricultural and Food Chemistry. 1984;**32**(3):447-453. DOI: 10.1021/jf00123a006

[68] Puri M, Sharma D, Barrow CJ. Enzyme-assisted extraction of bioactives from plants. Trends in Biotechnology. 2012;**30**(1):37-44. DOI: 10.1016/j.tibtech.2011.06.014

[69] Anis A, Anne SM. Discriminated release of phenolic substances from red wine grape skins (*Vitis vinifera* L.) by multicomponent enzymes treatment. Biochemical Engineering Journal. 2010; **49**(1):68-77. DOI: 10.1016/j.bej.2009.11.012

[70] Li BB, Smith B, Hossain MM. Extraction of phenolics from citrus peels: II. Enzyme-assisted extraction method. Separation and Purification Technology. 2005;**48**: 189-196. DOI: 10.1016/j.seppur.2005.07.019

[71] Pinelo M, Arnous A, Meyer AS. Upgrading of grape skins: Significance of plant cell-wall structural components and extraction techniques for phenol release. Trends in Food Science and Technology. 2006;**17**(11):579-590. DOI: 10.1016/j.tifs.2006.05.003

[72] Robledo A, Aguilera-Carbo A, Rodríguez R, Martinez JL, Garza Y, Aguilar CN. Ellagic acid production by *Aspergillus niger* in solid state fermentation of pomegranate residues. Journal of Industrial Microbiology and Biotechnology. 2008;**35**(6):507-513. DOI: 10.1007/s10295-008-0309-x

[73] Bhanja T, Rout S, Banerjee R, Bhattacharyy BC. Studies on the performance of a new bioreactor for improving antioxidant potential of rice. LWT - Food Science and Technology. 2008;**41**(8):1459-1465. DOI: 10.1016/j.lwt.2007.08.015

[74] Meyer AS. Enzyme technology for precision functional food ingredients processes. Annals of the New York Academy of Sciences. 2010;**1190**: 126-132. DOI: 10.1111/j.1749-6632.2009.05255.x

[75] Victor dF, Elisabete C, Nuno M. Study of carbohydrate influence on protein–tannin aggregation by nephelometry. Food Chemistry. 2003; **81**:503-509

[76] Luck G, Liao H, Murray NJ, Grimmer HR, Warminski EE, Willianson MP, et al. Polyphenols, astringency and proline-rich proteins. Phytochemistry. 1994;**37**(2):357-371. DOI: 10.1016/0031-9422(94)85061-5

[77] Gustavo HFS, Edvane BS, Bruna LS, Kêsia XFRS, Cláudia SAL. Influence of gamma radiation on the antimicrobial activity of crude extracts of *Anacardium occidentale* L., Anacardiaceae, rich in tannins. Revista Brasileira de Farmacognosia. 2011;**21**(3):444-449. DOI: 10.1590/S0102-695X2011005000045

[78] Qianqian H, Xiuli L, Guoqi Z, Tianming H, Yuxi W. Potential and challenges of tannins as an alternative to in-feed antibiotics for farm animal production. Animal Nutrition. 2018;**4**: 137-150. DOI: 10.1016/j.aninu.2017.09.004

Section 3

Applications in Agricuture and Industry

Chapter 5

Hydrolysable Tannins in Agriculture

Sergio Miele, Stefania Tegli, Carlos Garcia Izquierdo, Matteo Cerboneschi and Enrica Bargiacchi

Abstract

Hydrolysable tannins, water-extracted from sweet chestnut (*Castanea sativa* Mill.) (CHT) and membrane concentrated, have several effects as antioxidant, antimicrobial, and metal complexing agents. Some patents described their use as nitrogen release modulators and iron complexing agent to fight plant chlorosis and to control legume seed-borne disease and nitrosamine and mycotoxin during plant and food processing. Biostimulating activity of raw CHT, placed near seed or transplant seedlings, was assessed on early plant growth (starter effect) and found related to earlier production of a larger plant fine root mass, with greater P early uptake. Increased resistance to nematodes, with CHT applications on tobacco, was investigated. Recent process stream fractioning permitted to identify some CHT fractions with antimicrobial and antioxidant effects which were tested for their potential in promoting selected aspects of plant yield, quality, and protection and maintaining and improving feed and food quality during processing. EU Life+2013 Evergreen found a method of application of a CHT fraction to protect tobacco and carrot plants in nematode-infested fields. A protective effect of CHT on some bacterial diseases of olive tree and kiwi was disclosed. Environmental and soil toxicities were also investigated finding very low impacts and the possibility to reduce Cu use in agriculture.

Keywords: plant biostimulant, hydroculture, gall nematodes, TSNA, mycotoxins

1. Introduction

Polyphenols, a complex group of phytochemicals derived from phenylalanine, are characterized by an aromatic ring with a reactive hydroxyl group. They include phenolic acid derivatives, called hydrolysable tannins (HTs). They present a carbohydrate molecule (generally ∂-glucose), esterified at various levels with gallic or ellagic acids. These gallotannins and ellagitannins are hydrolyzed by weak acids or bases and are more prone to oxidation than condensed tannins (CTs) which are characteristics of oak wood and grapeseed extracts. Most HTs are typical of Mediterranean plants, but only tannins from water-extracted sweet chestnut (*Castanea sativa* Mill.) biomass (CHT) have been industrially exploited in agriculture as a corrective, a fertilizer chemistry modifier, and/or a biostimulant product, with a protective activity against nematodes and some microbial strains. Another important agricultural use, as a feed ingredient, is in expansion [1], while some technological applications in food processing and supplement production are just at their beginning [2, 3]. Other Mediterranean HTs "competitors" of CHT, i.e., those of myrtle (*Myrtus communis* L.) and pomegranate (*Punica granatum* L.), have found so far a much larger use in herbal medicine and supplement production [4].

A previous paper dealt with water-extracted CHT chemistry, in particular with the different fractions that can be separated by membrane concentration and reverse osmosis [5]. These fractions can be industrially mixed to maintain a fairly consistent composition in the final product, with better consistency of the agronomic results.

CHT have several remarkable effects as antioxidant, radical scavenging, antimicrobial, and metal complexing agents. Typically, tannins are renowned for their capacity to precipitate proteins, pectins, and cellulose. So they can inhibit many enzymes [6]. This is considered one of the major causes that reduces urea losses for volatilization and—in general—modifies nitrogen release in soluble forms.

Due to their fraction of non-tannins, and to the chemical structure of the tannins, they easily form complexes with several polyvalent ions [7]. Those with iron, manganese, and zinc are retained particularly interesting under the agronomic standpoint.

However, the single, most interesting effect of CHT on plants is related to their biostimulant effect on plant early rooting, which determines an increased resilience to abiotic stress (water and nutrient shortage in the soil) and some biotic stress also, e.g., soilborne disease and nematodes [8].

2. Effect of CHT in plant fertilization

2.1 Effect of CHT on water and fertilizer acidification

The first characteristic of CHT our research team has investigated, since 1999, was related to the acidifying effect of these tannins when applied to water and fertilizers. For organic farming, which relies almost exclusively on unacidified phosphate rock as a P_2O_5 source, a microgranulated (0.5–0.8 mm ø) 5.14.0–15 SO_3–28 C fertilizer, for local application at crop planting (rates: 35–50 kg/ha), was developed, made of phosphate rock, CHT, sulfur, and blood meal. In comparative tests, this fertilizer performed equally or better than ammonium phosphate, applied at much larger rates (150–200 kg/ha). In particular, co-granulation of 5–10% dry CHT with powdery phosphate rock increased 2.6–6 times the water and neutral ammonium citrate solubility [9].

For organic farming, CHT were used to acidify water used as a carrier of azadirachtin, a soil-applied natural insecticide-nematicide, which is stable only at acid water pH. This contributed to achieve more consistent results of azadirachtin vs. an ordinary treatment with citric acid.

2.2 Effect of CHT on nitrogen, iron, other microelements, and proposed fertilizer products

First studies on the inhibiting effect of purified tannins, both of sweet chestnut and Australian *Acacia* (*Acacia mearnsii* De Wild.), on ammonium sulfate nitrogen indicated they are not toxic to microbial growth; however, they reduce nitrification rate of some 20–30% at concentrations as low as 0.5–1.0% in the first weeks after application, and this effect remains for 12 weeks. This effect does not indicate a direct toxicity of tannins to nitrification bacteria [10]. The author of this research speculates also a reduction in soil-free ammonium nitrogen after ammonium application with CHT, due to the growth of heterotrophic microorganism on the carbon-rich substances (sugars, acids, etc.) usually present in commercial CHT.

Later, different mechanisms were proposed, inhibition of extracellular enzymes, iron deprivation, etc. [11], and more recently, this effect was explained on the basis of chemically combining tannins with extracellular enzymes and with an anti-infective action [12]. Interactions between CHT and chemistry of nitrogen forms were in part disclosed by a patent, dealing with the reduction of tobacco-specific nitrosamines by the use of field and/or post-harvest treatments [13].

An interesting point of view on this matter was carried out by the results of a paper dealing with condensed tannins of poplar species: small tannins (tetramer and smaller molecules) have a direct biological effect, as they act more often as substrata and sometimes as toxins, while highly polymerized tannins determine a reduction of net mineralization through substrate binding, with otherwise limited effects on overall C cycling or microbial communities directly [14]. Similar researches should be done also in the case of CHT.

Concerning the anti-urease activity, the so-called tannic acid (gallotannins) was found to determine a marked inhibition of urease activity (the inhibition constant was Ki = 0.040). The concentration of 0.1 mM reduces the urease activity of 72.4%. Apparently, the kinetic parameters (V_{max} and K_m) depend upon the concentration of "tannic acid," therefore suggesting the presence of a noncompetitive, pure inhibition mechanism. Soluble and insoluble tannin-urease complexes are formed both the longer the time of their contact and the higher the ratio "tannic acid to urease." The mechanism is related to the formation of reversible and nonreversible bonds with the substrate. Oxidative polymerization of tannins, which becomes evident by their browning, makes the enzyme progressively less accessible [15].

Low concentrations of CHT (0.8–1%) in a nitrogen-urea solution (18–20% N) determine a releasing curve which is comparable to that of a granular methylene urea. Lab tests carried out according to the Stanford and Smith incubation and leaching methodology, at equal N rate, indicated that a urea solution 20% with 1% CHT had a potential total nitrogen efficiency of 65–70% vs. 50% of urea (control), which releases completely in 28 days (**Figure 1**) [16].

Tests on corn and wheat consistently demonstrated that this technology is permitted to maintain yields with a reduction of the nitrogen rate up to 25%. In the case of wheat topdressing and topsoil applications on alkaline and subalkaline soils, reduction in volatilization losses represents the major mechanism of action for the increased efficiency.

Complexes of condensed tannins of *Quercus falcata*, *Salix sieboldiana*, and most of the Japanese conifers were studied as adsorbents of heavy metal ions, among which Cr^{6+}, Cd^{2+}, but also Cu^{2+}, and Fe^{2+} [18]. Later, *Larix* spp. condensed tannins

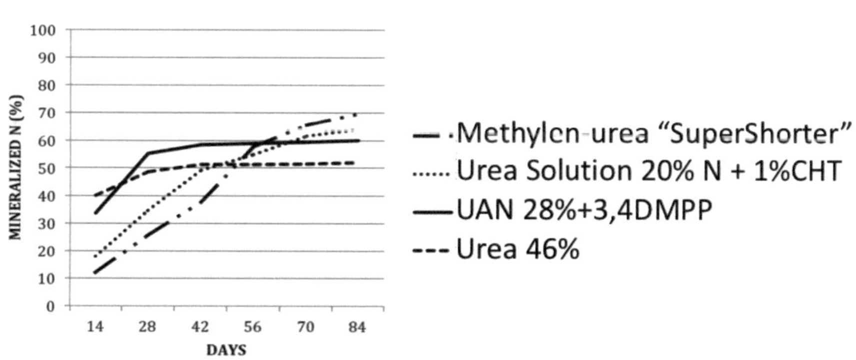

Figure 1.
Nitrogen mineralization determined as nitrates in the leachates after incubation in standard conditions (medium, temperature, moisture) according to Stanford and Smith method as modified by [17].

with iron, manganese, and zinc were prepared and applied to correct nutrient deficiency in some crops, e.g., apple orchards.

Iron chlorosis is one of the major problems both for fruit orchards and greenhouse crops. The use of mineral iron fertilizers, i.e., ferrous sulfate, has a low agronomic impact, due to the rapid oxidation to Fe^{3+} in the soil, while only the Fe^{2+} is readily available for plants. This is the reason why synthetic chelates and complexes have received such a deep interest over the last years.

CHT-iron complex fertilizers may represent a "green" alternative to synthetic chelates. In the formulated fertilizers, ¾ w/w of the total iron is as complex, according to the Italian Official Test method, which uses a cation-exchange resin [19]. Field tests were carried out mainly on pear, kiwi, and vineyard to treat iron chlorosis. On pear cv. Abate, on Ba29 rootstock, three applications in micro irrigation at 100 L/ha (6% Fe from Fe sulfate heptahydrate) gave comparable or better results than the ordinary treatment of Fe(EDDHA) at 12 kg/ha in three applications. A longer shelf life (+18 days, compared with the control, at room temperature) and better refrigerator preservability (−16% incidence of softening at 60 days from harvest, as compared with the control) were the main quality improvements [20].

The Italian Decree 2010/75 concerning domestic fertilizers, complementary to the European law 2003/2003, considers the following products with tannins:

- Two water/soil correctives: one liquid plant extract with tannins 13% and one solid plant extract with tannins, minimum 75%

- A solution, minimum 20% nitrogen, fertilizer, with 0.8% minimum tannins, pH 4–6

- An iron complex fertilizer with plant extracts with tannins, 6% total Fe, 4.8% complex Fe

Only later, with the Decree 2018-07-18 related to the production and labeling of corroborants, to potentiate plant defense, an integral extract of sweet chestnut with tannins was included in this group, with the only requirements to indicate the tannin percent content, the use of water extraction, and physical means to concentrate the product.

2.3 CHT as a biostimulant

The raw CHT extract has found application in transplanted crops as a starter treatment, to boost early plant growth, rooting, and phosphate uptake and enhance plant resistance to nematodes. Some recent data of 2017–2018 indicated that the use of tannins in the hydroculture medium, where transplant seedlings were grown, was positive for early rooting and growing after their field transplanting, and differences maintain until harvest.

During early characterization, some mixes higher in gallic acid (HGA) or lower in gallic acid (LGA) were compared. Both tannin formulations decreased tobacco actual infestation (root gall index) at 30 DAT, but only the HGA-mix significantly reduced the J2 population vs. the infested control. In fact, for this parameter, there were no differences between the LGA-mix treatment and the infested control. However, tobacco plants treated with this same LGA-mix produced an epigeal DM yield not significantly different from the HGA-mix treated plants. This seems to indicate that products low in gallic acid, such as the LGA-mix and most commercial CHT extracts, have a predominant biostimulant activity, while a higher concentration of free gallic acid is associated with an increased biocidal activity, in agreement with the previous literature [21].

A recent paper reported a large part of the activity on tannin-nematode carried out during the European Project Life+2013 Evergreen [22]. For the first time, one of the CHT fractions was formulated as a microgranulated fertilizer and a fully dispersible powder for application at low rates in efficient, localized repeated placements along each transplant row. These experiments confirm that CHT acts as a biostimulant on plant root systems, enhancing indirectly plant resistance to nematodes. This mechanism could positively affect CHT efficacy in the medium long term, because it determines a lower selective pressure on nematode population than more aggressive a.i. (active ingredients). It should also be noticed that this sustainable approach was demonstrated to be coupled with a null toxicity profile on model organisms and microorganisms by researchers of the same project Evergreen.

3. Potential and opportunities of tannins in the control of biotic plant diseases

The food demand is increasing worldwide for the constant growth of the global population, with a significant impact on natural resources, such as water, land, and biodiversity, that are already under pressure. Moreover, climate change is also further threatening food security, also by negatively affecting incidence, severity, and distribution of biotic plant diseases of plants. In this global scenario, where agriculture is pivotal as a source for both food commodities and income, more efficient, innovative, and sustainable production and control methods need to be urgently developed and adopted to prevent crop yield and quality losses. Other important challenges for future plant disease management are also to be able to preserve the environment, agroecosystems, and human health, as well as to reduce dependency from natural resources. In this regard, synthetic agrochemicals have been essential to determine a noteworthy increase in crop yields and food production during the last century, but their extensive use for the control of plant diseases resulted in an undeniable negative impact on the environment and ascertained risks for human health. In addition, the increasing demand for organic plant food has also inevitably determined the request for more environmentally friendly and safer pesticides, possibly from natural origin, to be used both in pre- and post-harvest disease management.

Safe and effective bioactive compounds useful as alternatives to synthetic pesticides for a sustainable plant disease control can be obtained by exploiting those secondary metabolites naturally produced by the plants also to provide protection against phytopathogens and pests, such as phenolic compounds. Plant phenolics form a class of chemically heterogeneous compounds, which include tannins, which occur in monocots, dicots, and ferns. Tannins were firstly classified into hydrolysable and condensed tannins; the former is absent in monocots, while their most recent classification is based on their structural characteristics and includes gallotannins, ellagitannins, complex tannins, and condensed tannins.

Tannins, traditionally used in the leather industries, have been widely investigated also for their beneficial effects on human health, including their antioxidant bioactivity and antimicrobial properties against some human pathogens. In the past, tannin-based extracts have been obtained from several medicinal plants, and used accordingly in traditional ethnomedicine of some countries, such as China, Japan, and Malaysia. In more recent times, tannins have been also studied to assess their potential as natural bactericides and fungicides in plant protection and management, to replace or reduce synthetic pesticides. Furthermore, the availability of green extraction and purification procedures to obtain raw or purified tannin extracts from agro-industrial waste makes these bioactive compounds highly attractive, contributing to the sustainability of agricultural practices in the frame of a circular economy [23].

The quali-quantitative yields obtained in the recovery of bioactive tannins are strictly related not just to the plant species used, as well as their specific parts or by-products and wastes, but also to the extraction procedures adopted. Microwave- and ultrasonic-assisted extraction of tannins are among the most innovative procedures adopted, to overcome the main limits of the traditional methods, by reducing extraction temperatures and times, as well as the amount of solvents used. As a general rule, tannins are not only soluble in water but also in several alcohols, ethyl acetate, and acetone, with differences in yield attributable to other factors such as the temperature and the extraction time and the ratio of liquid to plant solid matrix. These extracts, composed by several phytochemical constituents, are usually more active than the individually isolated compounds, because of synergistic effects [24]. In this frame, a clear and homogeneous standardization would be then desirable both for the extraction procedures adopted and for the *in vitro* and *in planta* tests to assess the antimicrobial effectiveness of tannin extracts in controlling biotic plant diseases [25]. Accordingly, the antimicrobial and anti-infectivity activities of standardised tannin extracts were deeply investigated and demonstrated in the EU LIFE 13 EVERGREEN project, coordinated by Prof. Stefania Tegli.

At last it is worth to mention that recently, formulations based on *Castanea sativa* and *Schinopsis lorentzii* tannins and on *Vitis vinifera* cane tannins have received a positive evaluation as "basic substances" for their use in plant protection by EFSA [26].

3.1 Tannins against plant pathogenic fungi

Fungi are important agents of pre- and post-harvest diseases of plants and of their products, causing heavy economic losses which can increase up to 50%, in developing countries and under highly severe conditions. Additionally, some phytopathogenic fungal species are known to produce mycotoxins, both under field conditions and mainly on harvested crops, causing considerable economic losses and high health risks for consumers.

To this concern, extracts from leaves of *Capsicum annuum* accession no. CGN 21526 (10 g plant material/100 ml solvents, to get a vacuum-dried powder at a final concentration of 50 mg/ml in water) have been proven effective against *Alternaria alternata* causing post-harvest infections on tomato [27]. The most abundant phenolic compound of all these extracts was gallic acid, whose concentration ranged from 23 to 64% of the total phenols. When added to the culture medium at the final concentrations of 5, 10, and 25 mg/ml, these extracts inhibited the mycelial growth *in vitro* from about 43 to 82% in comparison to the negative control, in a dose-dependent manner and with the extract obtained using water as solvent as the most active. The germination of *A. alternata* conidia was also inhibited by these *C. annuum* extracts, in the range 40–53% in comparison to the untreated control and with the extracts obtained using ethanol or ethyl acetate as solvents having the highest bioactivity. Similarly, soft rot caused by *A. alternata* cherry tomato fruits was reduced by the treatment (10 and 25 mg/ml) with these *C. annuum* extracts and with ethanol and ethyl acetate extracts as the most active.

Similarly, water and alcoholic extracts from pomegranate peel have been proven effective in controlling post-harvest rot and decay of fruits and crops, caused by several fungal phytopathogens such as *Penicillium* species, mycotoxigenic *Aspergillus*, *Botrytis cinerea*, and *Colletotrichum gloeosporioides*. Their bioactivity is also conserved when used in bioformulations with edible natural coatings, such as chitosan and alginate [28–30]. The antifungal properties of pomegranate peel are attributable to ellagitannins, such as punicalin, punicalagin, and ellagic acid, as well as gallotannins [31]. By comparing different extraction solvents, the recovery of bioactive tannins as well as anthocyanins was higher when using hydroalcoholic solutions at high concentrations, instead of just hot water. The 80% ethanol/water extract from pomegranate peel, concentrated by evaporation, was demonstrated highly effective to reduce the development of *Botrytis cinerea* rots on table grape and to control olive anthracnose caused by *Colletotrichum* spp. [32, 33]. An inhibitory effect was demonstrated for an ethanol extract of pomegranate peel for *A. alternata*, *Fusarium oxysporum*, *Phoma destructiva*, *Rhizoctonia solani*, and *Sclerotium rolfsii* [34]. Conflicting results have been obtained with a pomegranate peel water extract, amended to PDA medium. While at 8.60 and 17.20 mg/ml, the ability to inhibit *A. alternata* and *Fusarium* spp. was confirmed, and the antifungal activity against *Stemphylium botryosum* demonstrated for the first time, no decrease of the growth rates of *P. expansum* and *B. cinerea* was induced for this pomegranate peel water extract, as well as no effect found on *P. digitatum* [35]. The inhibitory activity was shown to be correlated with the concentration of total polyphenolics and in particular with punicalagins. However, according to the extraction procedure adopted, organic acids and other primary and secondary metabolites were known to be present in this water extract [36], which can account for the unexpected and conflicting results on its antifungal activity.

Tannin extracts from waste biomass of chestnut (*Castanea sativa*) have been shown to be highly promising for their potential use as natural fungicides in plant protection. They are obtained by solvents such as water and ethanol classified as generally recognized as safe (GRAS) to powdered dried chestnut burs. The inhibitory activity found on the growth of *A. alternata*, *F. solani*, and *B. cinerea* was dependent from the extraction procedure adopted as well as from the different sensitivity of the fungal species examined. However, it was found to be mainly attributable to ellagic acid, with EC50 values ranging from 13.33 to 112.64 μg/mL [37].

Condensed tannins, such as catechins, have been found in water extracts from chili (*Capsicum frutescens*) (8.50 mg/ml) and garlic (*Allium sativum*) (6.93 mg/ml) extracts, having strong antifungal activity *in vitro* against *C. gloeosporioides*, *Fusarium*, and *Phomopsis* spp. [38].

3.2 Tannins against plant pathogenic bacteria

The control of bacterial diseases of plants is extremely challenging because of the complex biological cycle of these phytopathogens, spanning from epiphytic and/or endophytic asymptomatic stages to survival into soil, water, or other wild host or nonhost plants, although their impact causes serious economic losses concerning yield and quality of a huge number of crops. Moreover, most of the management of bacterial diseases of plants still relies on the preventive use of copper compounds as bactericide, with a wide spectrum of well-known negative ecotoxicological consequences. In spite of that, studies concerning the search and the development of botanical extracts alternative to copper treatment against phytopathogenic bacteria are surprisingly less common than those for fungi and for bacteria pathogens on humans.

Condensed tannins based on flavan-3-olic units, such as catechins and epicatechins, have been shown to be the most abundant polyphenolic metabolites found in the water extracts of grape seeds and green tea. Grape seed polyphenolic extracts entirelies consist of several catechins and epicatechins with molecular weights ranging from 290 to 1170 Da and of free gallic acid, while in green tea extracts epigallocatechin gallate and epicatechin gallate represent 96% and 4% of the total tannins, respectively. By using an *in vitro* plant model system, both these tannin extracts were demonstrated to be able to give a statistical reduction of the hyperplastic symptoms produced by the inoculation of *Pseudomonas savastanoi* pv. *nerii* strain Psn23 on cuttings from 2-year-old twigs of *Nerium oleander*, in comparison to those untreated. In addition, a significant decrease of bacterial multiplication was observed on tannin-treated plants, which was comparable to the in planta growth of the Δ*hrpA* nonpathogenic mutant of Psn23 [39].

A strong antibacterial activity against the destructive causal agent of tomato bacterial wilt *Ralstonia solanacearum*, both *in vitro* and *in vivo*, was found for tannins extracted from *Sedum takesimense* and *Sapium baccatum* [40].

The profiling of the phenolic compounds present in young leaves of the two apple cultivars "Enterprise" and "Idared," highly resistant and highly susceptible to fire blight, respectively, was estimated and evaluated both before and after *E. amylovora* infection. According to this data, the activity of 13 selected phenolics was tested *in vitro* against *E. amylovora*, at 10, 50, and 100 mM in aqueous solution. Gallic acid was among the most effective to suppress the bacterial growth. Moreover, its efficacy was confirmed *in vivo*, by significantly limiting the development of disease and *E. amylovora* infection on pear fruitlet slices when applied as 100 mM aqueous solution.

3.3 Tannins and their mechanisms for plant disease control

The biological activity of tannins strongly depends from their highly variable chemical structure, and tannins basically can act as metal ion chelating and protein complexing agents and antioxidants, in addition to their well-known antimicrobial properties. However, in view of their potential application in plant protection, a deeper knowledge about their mode of action would be desirable.

Experiments carried on bacterial phytopathogen *Clavibacter michiganensis* with the ellagitannin HeT extracted from strawberry leaves demonstrated that its bactericide activity is related to a dose-dependent inhibition of the oxygen consumption rate and respiration, as a consequence of its interaction with cell membranes [41]. The absence of any toxicity was assessed for several tannins, such as epigallocatechin gallate and catechin, up to 1 μM by using as a target the membrane protein Ca^{2+}-ATPase from the sarcoplasmic reticulum (SR). SR belongs to the ubiquitous and highly conserved proteins of the so-called P-type ATPase family, whose members are present in the cellular membrane of any living organism and involved in numerous transport processes. Conversely, copper suppresses almost completely Ca^{2+}-ATPase activity at just 0.1 μM concentration Cu^{2+} ions [39].

An alcoholic extract obtained from the peel of pomegranate, and mainly containing tannins, was found active as resistance inducer. This extract elicits defense responses when applied to harvested citrus fruits, expressed as an increase in reactive oxygen species and in the expression of five genes which are pivotal during the activation of plant post-infectional defense [33].

Tannins have been also shown to possess noticeable inhibition abilities on some enzymatic activities strictly related to the virulence of phytopathogenic bacteria. The virulence of the *Dickeya chrysanthemi* is mainly dependent from its production and secretion of several cell wall-degrading enzymes, such as pectate lyases and proteases. Tannic and gallic acid are efficient to give a total inhibition of *D. chrysanthemi* pectate lyase at concentrations of 400 and 800 μg/ml, respectively [42]. At last, tannins can also negatively interfere with other bacterial systems which are essential for their pathogenicity and virulence on plants, such as the Type Three Secretion System and the Quorum Sensing, respectively [39, 43].

4. Polyphenol extracts to combat some bacterial diseases on kiwi crop: effect on soil biochemical functions

4.1 About soil quality

Soil is a natural resource that we must conserve and protect. In this sense, we must ensure that various soil properties (physical, chemical, biological, microbiological, and biochemical), capable of maintaining the quality, sustainability, and functionality of soils, respond to the soil protection criteria. Soil properties effected by the size, activity, and the composition of the microbial biomass included water holding capacity, infiltration rate, erodibility, aggregate stability, nutrient cycling, nutrient capacity, and soil organic matter content (soil function). Soil quality cannot yet be defined in quantitative terms; however, it should be possible eventually to define soil quality and sustainability quantitatively by the appropriate integration of specific quantitative terms, so that the effects of management on soil quality can be determined. Soil quality is a dynamic character, and many significant indicators must be sensitive to small changes in key soil properties. However, tools to detect the impact of changes in soil functionality are needed. Soil enzymes are extremely important in assessing the status or the conditions of the soil environment. This is because enzymes' and microorganisms' activity and biodiversity are related with the most important elements for soil sustainability and functionality (C, N, P, and S). Many extracellular enzymes are absorbed to, complexed with, or entrapped within soil clays and humics, and they may have a long-term stability.

The demand for biofertilizers is increasing since the last decade owing to its eco-friendly characteristics and a worldwide trend to reduce the reliance on chemically derived fertilizers. The Asia-Pacific shared approximately 34% of the total demand in 2011. European and Latin American countries are the leading consumers of biofertilizers, owing to stringent regulations imposed to chemical fertilizers which would eventually be replaced by biofertilizers.

4.2 About biopesticides

In the Evergreen Project, different experiments using polyphenol extracts as biopesticides were carried out. We show (only as an example) some results obtained on soils from a kiwi (*Actinida chinensis*) crop, where some polyphenol extracts were used as biopesticides. The bacteria (*Pseudomonas syringae actinidiae*) were used as pathogen agent, a vascular pathogen, whose most conspicuous symptom is the red-rusty exudation which covers bark tissues on trunks and twigs.

The polyphenol extracts used in this experiment were the following:

Form 1 (liquid): chestnut polyphenol 2%, olive polyphenol 1% in water (1:10).

Form 2 (liquid): chestnut polyphenol 1,5%, olive polyphenol 1%, and grape seeds 0.3% in water (1:10).

In addition, $CuSO_4$ has been used to compare a possible conventional treatment and another way to combat some pathogen microorganisms (biopesticides as polyphenol extracts). The total treatments in this assay were (1) control− (without bacteria), (2) control+ (with bacteria), (3) $CuSO_4$− ($CuSO_4$ without bacteria), (4) $CuSO_4$+ ($CuSO_4$ with bacteria), (5) form 1, and (6) form 2.

In this study, some biological and biochemical parameters measured on soil treated with polyphenol extracts have been shown. The use of biochemical parameters (soil enzyme activities) can be important to know the possible effect of polyphenols on the cycle of the important elements such as C, N, and P.

Application methods for polyphenols and pathogen:

a. Supply polyphenol (form 1/form 2) or $CuSO_4$ (100 c.c.) on soil next to the roots. Let it be absorbed during a week.

b. Spraying polyphenol or $CuSO_4$ solution on aerial part of the plant. Let it be absorbed (24 h).

c. Bacterial inoculation: remove leaves from each plant, petiole included (100% wounds done) along the stem, exposing the wound produced. Inoculate helping with a micropipette 10 μl of bacterial solution on the wound directly. Protect the wound with a film at least during 24 h, and then remove it.

d. Watering polyphenol treatments (100 c.c.) 7 and 15 days later from the bacterial inoculation with the corresponding liquid polyphenol (form 1 or form 2).

Results were the following.

4.2.1 Total organic carbon and total N

One of the most interesting parameters for soil quality is the organic carbon content, indicative of the organic matter content of the soil. The organic C induces fundamentally the productivity and fertility of the soil. Its presence in the soil is of great interest from two points of view: environmental (fixation of C in the soil) and agronomical (soil fertility).

In our experiment (**Table 1**), no significant differences were observed for organic C between the control soil and the soils treated with polyphenols. We know that polyphenols are organic products, and they should be implied in mineralization and humification processes; they could alter soil organic C. However, the addition of polyphenols to the soil did not alter organic C content in the soils studied. This confirms that the doses used for polyphenols in this study are not high enough to affect this type of parameter.

Nitrogen enhances plant growth, improves the quality of crops, and increases seed and fruit production. Nitrogen in the soil is usually supplied by decomposition of organic material, commercial fertilizers, and nitrogen-fixing bacteria. The desirable amount varies between crops; however, too much nitrogen can have adverse effects especially on the environment.

The differences found in soil total N in our experiment (**Table 1**) can be attributable to the variability of soil and our technical analyses; for this reason, it can be

	N total (g/100 g)	**C total (g/100 g)**	**Corg (g/100 g)**
Control–	0.210 a	8.153 a	3.516 a
Control+	0.606 b	9.716 ab	5.153 a
$CuSO_4$–	0.526 b	10.363 b	5.116 a
$CuSO_4$+	0.533 b	10.313 b	4.113 a
Form 1	0.533 b	11.066 b	4.546 a
Form 2	0.556 b	10.600 b	5.673 a

The same letter for each parameter indicates no significant differences between treatment (Tuckey's method, $p < 0.05$).

Table 1.
Total N, total C, and total organic carbon, measured in kiwi soils at the end of the experiment.

indicated that the variations observed in this parameter cannot probably be due to polyphenols' addition.

4.2.2 Enzymatic activities

Enzyme activities related to the cycle of elements (carbon, nitrogen, phosphorus, or sulfur) are of paramount importance in soil quality. Among these enzymes we propose the study of phosphatases, ureases, proteases, and different enzymes related to C cycle such as β-glucosidases. Indicators of the microbial population activity (dehydrogenase activity) will give an accurate notion of the impact of the addition of these products on microbial activity. For a general assessment of the functional and structural changes in microbial community, we have carried out several measurements based on soil enzymes.

Most enzymes found in the soil, in particular the hydrolases, are extracellular and have a great environmental interest. In addition, these extracellular enzymes may be free and exposed to rapid denaturation or immobilized together with mineral or organic colloids. Generally, those immobilized enzymes in mineral and/or organic colloid change in their status, nature, and properties (such as kinetics, stability and mobility of enzymes) and are less prone to proteolytic denaturalization, since they are physically and chemically associated with other surrounding chemical compounds.

4.2.3 Phosphatase

The agronomic and biotechnological importance of phosphatase is that it activates the transformation of organic P into inorganic forms of P available to plants. We have determined alkaline phosphatase since we have worked with basic soils (**Table 2**). Phosphatases are inhibited by inorganic P, the final product of their enzymatic reaction. This is due to a feedback inhibition, so phosphatases are synthesized only when available P is deficient.

In our study, no statistical differences between treatments were appreciated in this enzyme activity. This indicates that the addition to the soils of polyphenol products does not change the phosphorous cycle in the soil. Some differences in this enzyme activity were noted throughout the experimental period, greater phosphatase activity being detected at the start than at the end of the experiment. This fact could be due to P mobilization from organic to inorganic forms, in order to make it available to plants. The P cycle, studied by phosphomonoesterase activity, seems not to be affected by the polyphenol addition to the soil since little differences were observed as regards phosphatase activity between the soils treated with polyphenols

Phosphatase activity (μmol PNF $h^{-1} g^{-1}$ soil)	T0	Tf
Control−	6.779 b	4.028 a
Control+	7.118 b	5.197 a
$CuSO_4$−	6.072 ab	5.373 a
$CuSO_4$+	6.206 ab	5.560 a
Form 1	6.391 ab	5.886 a
Form 2	5.349 a	5.324 a

The same letter for each parameter indicates no significant differences between treatments (Tuckey's method, $p < 0.05$).

Table 2.
Evolution of soil alkaline phosphatase activity in kiwi soils (initial, To, and final Tf).

and the control. Anyway, a slight increase in phosphatase activity was observed when polyphenols were introduced into the soil.

A negative effect on soil phosphatase was observed when $CuSO_4$ is used as pesticide. It indicates that this conventional treatment can affect to P cycle in the soil.

4.2.4 β-Glucosidase

β-Glucosidase is a hydrolase which intervenes in the C cycle, acting especially in the hydrolysis of the β-glucoside bonds of long carbohydrate chains. The hydrolysis of these substrates plays an important role in the attainment of energy from the soil by microorganisms.

C cycle linked to β-glucosidase activity was not affected by the utilization of polyphenols, as shown in **Table 3**. The activity of this enzyme did not change with time in a significant way, similar activity values being observed at the start and end of the experimental period. Both polyphenols directly (form A and B and also the Cu salt introduction in soil) and their impact on the soil biota do not affect the carbon cycle.

Urease activity. Urease catalyzes the hydrolysis of urea or ureic-type substrate to give carbon dioxide and ammonia as reaction products. This term includes all those hydrolases capable of acting on the C-N (non-peptide) bonds of linear amides. They are extracellular enzymes.

At the start of the experiment, some changes in urease activity were observed in the soils treated with polyphenols with respect to the control (**Table 4**). This is indicative that the N cycle is influenced by polyphenols. The soils treated with $CuSO_4$ showed the lowest values of urease activity; it could be due to the heavy metal incidence (Cu) or to the increase of soil salinity. At the end of the experiment, urease activity values were also lower when $CuSO_4$ was used, but the differences were not statistically significant. For this reason, we can say that the microbial populations that synthesize urease do not undergo to experiment changes, and consequently, the N cycle does not show difference between soils.

D-hydrogenase. The biological oxidation of organic compounds occurs by means of dehydrogenation processes, in which intracellular enzymes called dehydrogenases take part. The dehydrogenation activity in soils is determined by different dehydrogenase systems, which are characterized by their high substrate specificity. All these systems are an integral part of the microorganisms; indeed, dehydrogenase activity has been proposed as an indicator of soil microbiological activity and biomass.

Dehydrogenase activity is intracellular and detects the set of cells capable of being activated against various situations; soil samples toward the initial of the

β-glucosidase activity (μmol PNF $h^{-1} g^{-1}$ soil)	TO	Tf
Control–	0.480 a	0.524 a
Control +	0.527 a	0.531 a
$CuSO_4$–	0.527 a	0.475 a
$CuSO_4$+	0.563 a	0.472 a
Form 1	0.544 a	0.620 a
Form 2	0.549 a	0.573 a

The same letter for each parameter indicates no significant differences between treatments (Tuckey's method, $p < 0.05$).

Table 3.
Evolution of soil β-glucosidase activity in kiwi soils (initial, To, and final Tf).

Urease (μg INTF h^{-1} g^{-1} soil)	TO	Tf
Control−	0.200 b	0.931 a
Control+	0.246 c	0.945 a
$CuSO_4$−	0.141 a	0.758 a
$CuSO_4$+	0.180 ab	0.726a
Form 1	0.364 d	1.008 a
Form 2	0.329 d	0.806 a

The same letter for each parameter indicates no significant differences between treatments (Tuckey's method, $p < 0.05$).

Table 4.
Evolution of soil urease activity in kiwi soils (initial, To, and final Tf).

D-hydrogenase activity (μg INTF h^{-1} g^{-1} soil)	TO	Tf
Control−	1.619 a	5.355 b
Control+	1.595 a	5.147 b
$CuSO_4$−	1.391 a	2.972 a
$CuSO_4$+	1.559 a	4.719 b
Form 1	1.174 a	5.274 b
Form 2	1.487 a	5.380 b

The same letter for each parameter indicates no significant differences between treatments (Tuckey's method, $p < 0.05$).

Table 5.
Evolution of soil d-hydrogenase activity in kiwi soils (initial, To, and final Tf).

experiment showed no changes for this oxidoreductase enzyme. Only at the end of the experiment, the soils treated with $CuSO_4$ showed a decrease in the activity of this enzyme. This different behavior between the start and the end of the experiment could be due to the fact that at the start of the experiment, the Cu salt has no time to act, and some more time needed to the effect of the salt on the enzyme activity is noted (**Table 5**).

The main conclusion obtained from the results is that the use of polyphenols such as those prepared in the Evergreen project can be regarded as positive, since it is able to prevent bacterial diseases on crops such as kiwi. Our results indicate that the polyphenols used can be considered as biopesticides. For example, we can indicate that in agriculture, Cu is a metal widely used as a pesticide; however, the accumulation of this metal in soils can become harmful to the quality of that soil. Therefore, the possibility of having alternatives such as polyphenols, capable of acting against certain pathogenic microorganisms, has paramount agronomic and environmental interests. We think that the use of polyphenols should be studied at the level of management. The application of these compounds (the time of application if application should be repeated, if they can be used as preventive treatment, etc.) should be studied, particularly for soil biological properties.

Acknowledgements

The Authors are grateful for financial support for publishing this chapter to Saviolife Co. of Saviola Holding Group, Viadana (Italy).

Author details

Sergio Miele[1*], Stefania Tegli[2], Carlos Garcia Izquierdo[3], Matteo Cerboneschi[2] and Enrica Bargiacchi[1]

1 Consortium INSTM, Firenze, Italy

2 Department of Agriculture, Food, Environment and Forestry (DAGRI), Plant Pathology and Entomology Section, University of Florence, Firenze, Italy

3 Centro de Edafología y Biología Aplicada del Segura (CEBAS-CSIC), Campus Universitario de Espinardo, Murcia, Spain

*Address all correspondence to: smiele.agr@instm.it

References

[1] Antongiovanni M, Buccioni A, Minieri S, Rapaccini S. Effetti dell'estratto di tannino idrolizzabile di castagno sul bilancio azotato di pecore, suini e broiler [CHT extract effect on Nitrogen balance of sheep, swine, and broiler]. In: Proceedings of Round-up "I tannini di castagno ed i loro molteplici impieghi" [Sweet chestnut tannins and their numerous uses], Accademia dei Georgofili, Firenze, March 22, 2012

[2] Bargiacchi E, Bellotti P, Pinelli P, Costa G, Miele S, Romani A, et al. Use of chestnut tannin extract as anti-oxidant, anti-microbial additive, and to reduce nitrosamines and mycotoxins. EP 15154069.7-1357. 2015-04-01 (priority: 2014-02-07 MI20140177)

[3] Kastania. 2019. Available from: http://www.kastania.it/en/ [Accessed: 11-02-2019]

[4] Romani A, Campo M. Caratterizzazione chimica dei tannini idrolizzabili di Castagno e da altre specie vegetali [Chemical characterization of HTs of sweet chestnut and other plant species]. In: Proceedings of Round-up "I tannini di castagno ed i loro molteplici impieghi" [Sweet chestnut tannins and their numerous uses], Accademia dei Georgofili, Firenze, March 22, 2012

[5] Bargiacchi E, Campo M, Romani A, Milli G, Miele S. Hydrolysable tannins from sweet chestnut (*Castanea sativa* Mill.) to improve tobacco and food/feed quality. Note 1: Fraction characterization, and Tobacco biostimulant effect for gall-nematode resistance. In: Toselli M et al. editors. AIMS Agriculture and Food, Special edition "Use of biostimulant in agriculture to improve plant performance and crop quality". Available from: http://www.aimspress.com/article/10.3934/agrfood.2017.3.324

[6] Hagerman AE. The Tannin Handbook. 3rd ed. Ohio University; 2011. Available from: www.users.miamioh.edu/hagermae

[7] Rao MA, Gianfreda L. Properties of acid phosphatase-tannic acid complexes formed in the presence of Fe and Mn. Soil Biology and Biochemistry. 2000;**32**:1921-1926

[8] Bargiacchi E, Costa G, Miele S, Magni S. Use of natural extracts of tannin and non-tannin materials for improving soil fertility and providing a starter effect on cultivations. 2010. EP2345628, priority IT2010MI00040 20100115

[9] Bargiacchi E, Miele S. Fertilizing composition containing phosphorite and tannins. 2000. EP1097912, priority IT1999MI02296 19991103

[10] Basaraba J. Effects of vegetable tannins on glucose oxidation by various microorganisms. Canadian Journal of Microbiology. 1966;**12**:787-794. DOI: 10.1139/m66-106

[11] Scalbert A. Antimicrobial properties of tannins. Phytochemistry. 1991;**30**:3875-3883

[12] Tegli S, Cerboneschi M, Biancalani C, Calamai S, Bini L. Next-generation ecofriendly control of gram negative plant pathogenic bacteria: virulence inhibiting peptides and polyphenols from non-food plant biomass. Journal of Plant Pathology. 2017;**99**:62-62. ISSN: 2239-7264

[13] Miele S, Romani A, Bargiacchi E, Campo M. Method for making low nitrosamine contents tobacco. 2014. WO2014024020, priority IT2012MI0141920120808

[14] Fierer N, Schimel JP, Cates RG, Zou J. Influence of balsam poplar

tannin fractions on carbon and nitrogen dynamics in Alaskan taiga floodplain soils. Soil Biology and Biochemistry. 2001;**33**:1827-1839. PII: S0038-0717.01)00111-0

[15] Gianfreda L, DeCristofaro A, Rao MA, Violante A. Kinetic behavior of synthetic organo- and organo-mineral-urease complexes. Soil Science Society of America Journal. 1995;**59**:811-815

[16] Miele S, Bargiacchi E, Tannini di castagno ad uso agricolo e industriale. In: Romani A, et al., editors. Estratti naturali da piante medicinali e tessili-tintorie. Caratterizzazione ed usi innovativi di ortica, daphne, lavanda e tannini di castagno. Resoconto del progetto PRIN 2008. Prato: Aracne editrice; 2014. pp. 380. ISBN: 978-88-548-6724-6

[17] Magni S, Foschi L, Piccotino D, Miele S. Nitrogen availability of different slow release fertilizers as determined by incubation in a sand based growing medium. In: Proc. European Turfgrass Society Congress, Pisa, May 19-20. 2008

[18] Ohara S. Chemistry and utilization of condensed tannins from tree barks. JARQ. 1994;**28**:70-78

[19] EN 13366 (2001) method. www.store.uni.com/catalogo/index.php/en-13366-2001.html

[20] Bargiacchi E, Costa G, Della Croce C, Foschi L, Pampana S, Miele S, et al. Method for obtaining fertilizers using organic extracts from aqueous wood leaching processes. 2014. EP1464635, priority IT2003MI00640 20030331

[21] Bargiacchi E, Miele S, Romani A, Campo M. Biostimulant activity of hydrolyzable tannins from sweet chestnut (*Castanea sativa* Mill.). Paper presented to the 1st World Congress on Biostimulants, Strasbourg (France), Nov. 26-29, 2012, Proc. Acta Horticulturae. 2013;**1009**:111-115

[22] Bargiacchi E, Campo M, Romani A, Milli G, Miele S. Hydrolysable tannins from sweet chestnut (*Castanea sativa* Mill.) to improve tobacco and food/feed quality. Note 1: Fraction characterization, and Tobacco biostimulant effect for gall-nematode resistance. In: Toselli M et al. editors. AIMS Agriculture and Food, Special edition "Use of biostimulant in agriculture to improve plant performance and crop quality". 2017. Available from: http://www.aimspress.com/article/10.3934/agrfood.2017.3.324

[23] Zuin VG, Ramin LZ. Green and sustainable separation of natural products from agro-industrial waste: Challenges, potentialities, and perspectives on emerging approaches. Topics in Current Chemistry. 2018;**376**(1):3. DOI: 10.1007/s41061-017-0182-z

[24] Akhtar S, Ismail T, Fraternale D, Sestili P. Pomegranate peel and peel extracts: chemistry and food features. Food Chemistry. 2015;**174**:417-425. DOI: 10.1016/j.foodchem.2014.11.035

[25] Vu TT, Kim H, Tran VK, Vu HD, Hoang TX. Antibacterial activity of tannins isolated from *Sapium baccatum* extract and use for control of tomato bacterial wilt. PLoS One. 2017;**12**(7):e0181499. DOI: 10.1371/journal.pone.0181499

[26] Outcome of the consultation with Member States and EFSA on the basic substance application for Castanea and Schinopsis tannins for use in plant protection as bactericide, fungicide and nematicide. European Food Safety Authority (EFSA). 2018. DOI: 10.2903/sp.efsa.2018.EN-1363

[27] Pane C, Zaccardelli M. Evaluation of Bacillus strains isolated from

solanaceous phylloplane for biocontrol of Alternaria early blight of tomato. Biological Control. 2015;**84**:11-18. DOI: 10.1016/j.biocontrol.2015.01.005

[28] Da Cruz Cabral L, Fernandez Pinto V, Patriarca A. Application of plant derived compounds to control fungal spoliage and mycotoxin production in foods. International Journal of Food Microbiology. 2013;**166**:1e14. DOI: 10.1016/j.ijfoodmicro.2013.05.026

[29] Tayel AA, Salem MF, El-Tras WF, Brimer L. Exploration of Islamic medicine plant extracts as powerful antifungals for the prevention of mycotoxigenic Aspergilli growth in organic silage. Journal of the Science of Food and Agriculture. 2011;**91**:2160-2165. DOI: 10.1002/jsfa.4433

[30] Nair MS, Saxena A, Kaur C. Characterization and antifungal activity of pomegranate peel extract and its use in polysaccharide-based edible coatings to extend the shelf-life of capsicum (*Capsicum annuum* L.). Food and Bioprocess Technology. 2018;**11**:1317-1327. DOI: 10.1007/s11947-018-2101-x

[31] Sneha Nair M, Saxena A, Charanjit K. Effect of chitosan and alginate based coatings enriched with pomegranate peel extract to extend the postharvest quality of guava (*Psidium guajava* L.). Food Chemistry. 2017;**240**. DOI: 10.1016/j.foodchem.2017.07.122

[32] Mirabedini A, Foroughi J, Romeo T, Wallace G. Development and Characterization of Novel Hybrid Hydrogel Fibers. Macromolecular Materials and Engineering. 2015;**300**:1217-1225. DOI: 10.1002/mame.201500152

[33] Pangallo S, Nicosia MGLD, Agosteo GE, Abdelfattah A, Romeo FV, Cacciola SO, et al. Evaluation of a pomegranate peel extract as an alternative means to control olive anthracnose. Phytopathology. 2017;**107**(12):1462-1467. DOI: 10.1094/PHYTO-04-17-0133-R

[34] Al-Askar AA. *In vitro* antifungal activity of three Saudi plant extracts against some phytopathogenic fungi. Journal of Plant Protection Research. 2012;**52**(4):458-462. DOI: 10.2478/v10045-012-0073-9

[35] Glazer I, Masaphy S, Marciano P, Bar-ilan I, Holland D, Kerem Z, et al. Partial Identification of antifungal compounds from *Punica granatum* peel extracts. Journal of Agricultural and Food Chemistry. 2012;**60**:4841-4848. DOI: 10.1021/jf300330y

[36] Dafny-Yalin M, Glazer I, Bar-Ilan I, Kerem Z, Holland D, Amir R. Color, sugars and organic acids composition in aril juices and peel homogenates prepared from different pomegranate accessions. Journal of Agricultural and Food Chemistry. 2010;**58**:4342-4352. DOI: 10.1021/jf904337t

[37] Esposito T, Celano R, Pane C, Piccinelli AL, Sansone F, Picerno P, et al. Chestnut (*Castanea sativa* Miller.) burs extracts and functional compounds: UHPLC-UV-HRMS profiling, antioxidant activity, and inhibitory effects on phytopathogenic fungi. Molecules; 2019;**24**(2). pii: E302. DOI: 10.3390/molecules24020302

[38] Ogbebor NO, Adekunle AT, Enobakhare DA. Inhibition of *Colletotrichum gloeosporioides* (Penz) Sac. causal organism of rubber (Hevea brasiliensis Muell. Arg.) leaf spot using plant extracts. African Journal of Biotechnology. 2007;**6**:213-218

[39] Biancalani C, Cerboneschi M, Tadini-Buoninsegni F, Campo M, Scardigli A, Romani A, et al. Global analysis of type three secretion system and quorum sensing inhibition of pseudomonas savastanoi by polyphenols extracts from vegetable residues. PLoS

One. 2016;**11**(9):e0163357. DOI: 10.1371/journal.pone.0163357

[40] Vu TT, Kim JC, Choi YH, Choi GJ, Jang KS, Choi TH, et al. Effect of gallotannins derived from *Sedum takesimense* on tomato bacterial wilt. Plant Disease. 2013;**97**(12):1593-1598. DOI: 10.1094/PDIS-04-13-0350-RE

[41] Martos GG, Mamani MA, Pedro PFO, Néstor A, Katz E, Castagnaro AP, et al. Ellagitannin HeT obtained from strawberry leaves is oxidized by bacterial membranes and inhibits the respiratory chain. FEBS Open Bio. 2017;**8**:211-218. DOI: 10.1002/2211-5463.12361

[42] Zaidi–Yahiaoui R, Zaidi F, Bessai AA. Influence of gallic and tannic acids on enzymatic activity and growth of *Pectobacterium chrysanthemi*. African Journal of Biotechnology. 2008;**7**(4):482-486

[43] Truchado P, Tomas-Barberan FA, Larrosa M, et al. Food phytochemicals act as Quorum Sensing inhibitors reducing production and/or degrading autoinducers of Yersinia enterocolitica and *Erwinia carotovora*. Food Control. 2012;**24**:78-85

Chapter 6

Applications of Tannin Resin Adhesives in the Wood Industry

Xiaojian Zhou and Guanben Du

Abstract

Tannin is extracted from natural sustainable materials. It is widely used to prepare tannin resin adhesives owing to its naturally occurring phenolic structure. This chapter aims to introduce the resources and structures of tannin, existing reactions that are involved in the synthesis of tannin resin, and the applications of tannin resin adhesives in the wood industry. Additionally, the advancements in the research based on the use of tannin resins in manufacturing plywood, particleboard, wood preservation, decoration paper impregnation, structural glulam, impregnated fibers, and other wooden products are reviewed. Herein, the main limitations encountered during the application of tannin resin adhesives and the future key research points are identified. Finally, the potential applications of tannin resin adhesives in the wood industry have been discussed.

Keywords: tannins, resins, adhesives, wood industry, applications

1. Introduction

The use of adhesives dates back to approximately 3000 years ago. Several types of adhesives based on specific applications have been developed, particularly for the manufacturing of wood and paper products, among other products. Therefore, thousands of adhesive products have been developed. Factors that affect the selection of the adhesives are cost, assembly process, bonding strength, and durability.

The fabrication of wood-based panel products involves a "preparation and recombination of wood unit" process wherein wood adhesives play a crucial role. Adhesives play a vital role in wood processing because their quality has a direct impact on the performance of the final wood product.

Synthetic and natural resins are the most commonly used adhesives in the wood industry. Some examples of synthetic resins are urea-formaldehyde resin; phenolic resin; melamine formaldehyde resin; and copolycondensation resin, which include phenol-urea-formaldehyde resin (PUF) and melamine–urea-formaldehyde resin (MUF). Some examples of natural resins are soy protein adhesive, tannin resin, lignin adhesive, and starch adhesive.

Although synthetic resin has high weathering resistance and mechanical strength, its raw materials are derived from nonrenewable petrochemical products that are volatile and expensive. Additionally, these products emit formaldehyde, which is toxic and carcinogenic.

The awareness of environmental protection and personal health has been emphasized in recent years. Therefore, natural resins with renewable resources as the main materials have attracted considerable amount of attention. Research and application of the tannin resin have been highly successful in some countries because its phenolic structure enables its use as adhesives and as a partial or complete substitute for phenols in adhesives. This chapter provides a comprehensive discussion of the situation of the existing tannin resources, reaction mechanisms involved in the synthesis of tanning resins, and general application of tannin resins in the wood industry. This information could provide ideas for the scholars and broaden the application scope of tannin resins in the wood industry.

The production of tannin for leather manufacturing peaked immediately after World War II and has progressively declined. Tannin adhesives were first successfully commercialized in South Africa in the early 1970s. Subsequently, mimosa tannin adhesives were used instead of synthetic phenolic adhesives to manufacture particleboard and plywood for external and marine applications. Tannin resin adhesives have been used in Australia, Zimbabwe, Chile, Argentina, Brazil, and New Zealand [1].

2. Tannin resources

Tannins are extracted from agroforestry biomaterials, such as wood, bark, leaves, and fruits, by the water extraction method. Tannins can be categorized as hydrolyzable tannin or condensed polyflavonoid tannin. The latter is one of the main objects of wood adhesive research and accounts for 90% of the global tannin output. The annual industrial output of tannin reaches up to 200,000 tons.

The distribution of tannin resources in the world has regional characteristics. For example, black wattle tannin is mainly manufactured in Brazil, South Africa, India, and other countries. Quebracho tannin is mainly manufactured in Argentina. Chestnut tannin is mainly manufactured in Italy and Slovenia. Pine bark tannin is mainly manufactured in Chile and Turkey. Oak tannin is mainly manufactured in Poland. Tannin from grape residues, such as skins and seeds, is mainly manufactured in France. In China, tannin is mainly synthesized from larch, poplar, and acacia bark.

3. Tannin structures

Hydrolyzable tannin comprises different types of unit structures, including gallic, digallic, and ellagic acids (see **Figure 1**), as well as sugar esters, which usually exist in the form of glucose [2, 3].

Condensed tannin comprises monoflavonoids or flavonoid units that have undergone various degrees of polymerization. These units are associated with their precursors, such as flavanes-3-ol and flavanes-3,4-diol, among other flavonoids [4, 5]. Each flavonoid contains two types of phenolic nuclei, which are A- and B-ring, as shown in **Figure 2**. The A-ring includes resorcinol and phloroglucinol, whereas the B-ring includes pyrogallol and catechol, among other rare phenols. The A-rings of different tannins possess different chemical structures. The A-rings of tannins extracted from mimosa/wattle, quebracho, Douglas fir, and spruce include resorcinol, whereas those of pine include phloroglucinol.

The main polyphenolic pattern is represented using flavonoid analogs that are based on the resorcinol A-ring and pyrogallol B-ring (I type in **Figure 3**). This unit structure accounts for 70% of tannin. Unit structure II constitutes 25% of tannin and comprises a resorcinol A-ring and catechol B-ring (II type in **Figure 3**). The remaining 5% is a mixture of phloroglucinol-pyrogallol (III type in **Figure 3**) and phloroglucinol-catechol (IV type in **Figure 3**) flavonoids. These four patterns constitute 65–80% of mimosa bark extract. The remaining components are non-tannins,

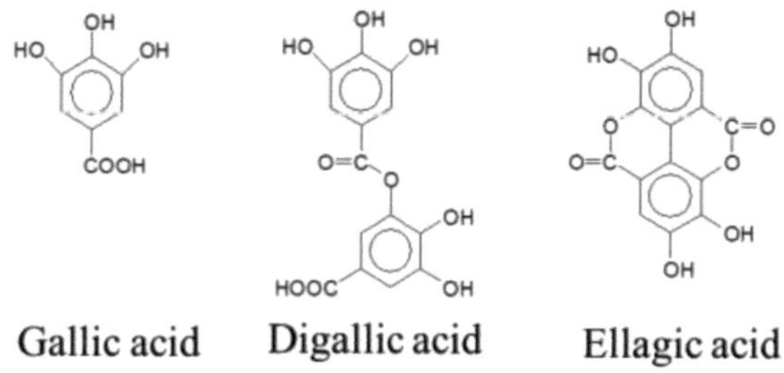

Figure 1.
Unit structures of hydrolyzable tannin.

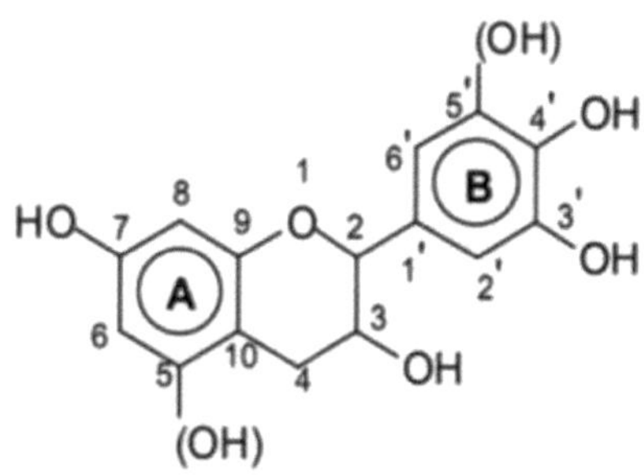

Figure 2.
Main structure of condensed tannin.

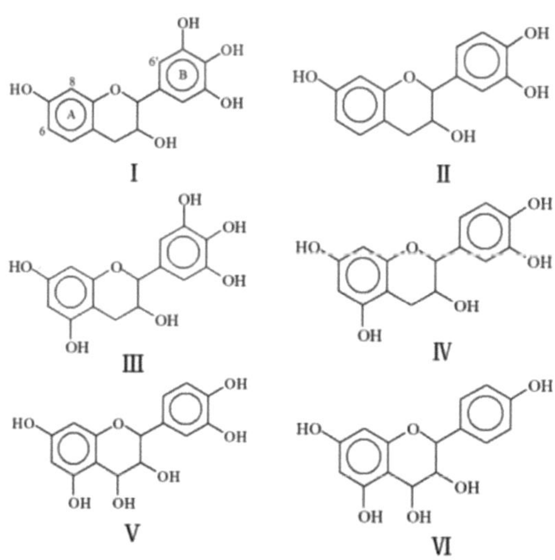

Figure 3.
Main units of condensed tannin.

which are simple carbohydrates, hydrocolloid gums, and nitrogen compounds, i.e., amino and imino acids. Gums and pectins are the most important components of tannins and have a significant effect on the viscosity of the extract despite their low concentration, i.e., 3–6%. These non-tannin substances can attenuate wood failure and can decrease the water resistance of glued products.

Pine tannin mainly presents two patterns: one is represented by phloroglucinol A-ring and catechol B-ring structures (V type in **Figure 3**) and the other is represented by phloroglucinol A-ring and phenol B-ring structures (VI type in **Figure 3**).

Flavonoid units can be bound through their 4,6- and/or 4,8-linkages to form polyflavonoids. Wattle-extracted tannin comprises 4–5 flavonoid units joined together through 4,6-linkages. Each unit of wattle-extracted tannin has an average mass number of 1250. The average mass number of quebracho tannin and pine is 1784 and approximately 4300, respectively. Pine tannin is phloroglucinolic in nature and its flavonoid units are joined together through 4,8-interflavonoid linkages. Linear polymeric tannins have only 4,6- (V) or 4,8-linkages (VI). However, 4,6- and 4,8-linkages may simultaneously exist in the presence of resorcinolic and phloroglucinolic A-rings. This phenomenon results in the synthesis of angular rather than linear polymeric tannins (VII). Matrix-assisted laser desorption/ionization time-of-flight revealed that mimosa tannin is highly branched owing to the presence of high proportions of angular units in its structure. On the contrary, quebracho tannin is almost completely linear. These structural differences contribute to the considerable differences in the viscosity of tannin water solutions [6].

4. Synthesis of tannin resin adhesives

The low reactivity of hydrolyzable tannins with formaldehyde limits their application in the wood industry, which can be attributed to their simple phenolic structures (**Figure 1**).

Tannin extracts usually contain sugars and gums, which are not involved in the synthesis of resin adhesives. Commercially available tannin extracts from black wattle and hardwood typically contain 70–80% of natural phenolic polymers, whereas those obtained from pine contain only 50–60% of natural phenolic polymers. Sugar dilutes the actual solid content, thus affecting the final properties of resins. Gum considerably affects the strength of the resin and water resistance of the adhesive. Due to the presence of non-tannin components, unmodified tannin adhesive is unsuitable for the production of wood products with high requirements. Therefore, tannin adhesives must be modified.

Normally, the viscosity of tannin resin adhesives is higher than that of synthetic resins at the same concentration due to (1) the presence of high molecular weight tannins in the extract and (2) the existence of hydrogen bonding and electrostatic interactions between tannin and tannin, tannin and gum, and gum and gum. Effective methods for decreasing the viscosity of tannin extracts in aqueous solutions include the following: (1) acid or alkaline hydrolysis of high molecular weight carbohydrates, e.g., with acetic anhydride, maleic acid anhydride, or NaOH [7, 8]; (2) addition of small amounts of hydrogen bond breakers (e.g., 3% urea based on the solid content of the extract); and (3) destruction of heterocyclic ether in tannin molecules through sulfite or bisulfite treatment.

4.1 Reaction of tannin with aldehyde

Tannin being phenolic in nature undergoes the same alkali- or acid-catalyzed reaction with formaldehyde experienced by phenols. Alkali-catalyzed reactions are predominantly used in industrial applications. Nucleophilic centers on the A-ring of any flavonoid unit tend to be more reactive than those on the B-ring. Thus, the reaction for inducing polymerization between formaldehyde and tannin mainly occurs on the A-ring through methylene bridge linkages. The A-ring of the condensed tannin molecules contains flavonoid units that possess one highly reactive nucleophilic center each. The reactivity of the resorcinol A-ring (e.g., wattle) toward formaldehyde is comparable with that of resorcinol. On the contrary, the phloroglucinol A-ring (e.g., pine) behaves as phloroglucinol. Pyrogallol or the catechol B-ring are unreactive and may only be activated via anion formation at a relatively high pH [9, 10]. Hence, the B-ring does not participate in polymerization except at a high pH (pH = 10). However, the reactivity between the A-ring and formaldehyde influences pot life because it is too fast to control.

In general, only the A-ring structure participates in crosslinking to build networks in tannin resin adhesives (**Figure 4**). However, owing to their size and shape, tannin molecules become immobile at low levels of condensation with formaldehyde. Thus, a large distance between the available reactive sites for further methylene bridge formation results in the incomplete polymerization of tannin resin adhesives. Incomplete polymerization, in turn, results in the formation of weak and brittle adhesives. Bridging agents with long molecules, such as phenolic and amino-plastic resins [10, 11], have been used to overcome this limitation by bridging the distances that are too large for interflavonoid methylene to bridge.

Catechol and catecholic B-ring do not react with formaldehyde at a pH value less than 10. Adding zinc acetate to the reaction mixture induces the B-ring to react with formaldehyde at low pH values, the optimum pH being in the range of 4.5–5.5, as shown by the high amount of formaldehyde being consumed. This finding implies that the further crosslinking of the tannin-formaldehyde network could be achieved through the participation of the B-ring in the reaction in the presence of zinc acetate. Strength can be improved through the addition of zinc acetate at economically acceptable levels (5–10% in resin solids). Nevertheless, improved strength is not comparable with the strength of fortified tannin resin.

Crosslinking is sometimes performed through the addition of isocyanate. The highly reactive diphenylmethane diisocyanate (MDI) can be used to assist the participation of B-ring in the crosslinking reaction [12]. Additionally, the reaction between polymeric diphenylmethane diisocyanate (pMDI) and carbohydrates or hydrocolloid gums can help in increasing the bonding strength of wood products. The reaction rate of wattle and pine tannins with formaldehyde is slowest in the pH range of 4.0–4.5 and 3.3–3.9, respectively.

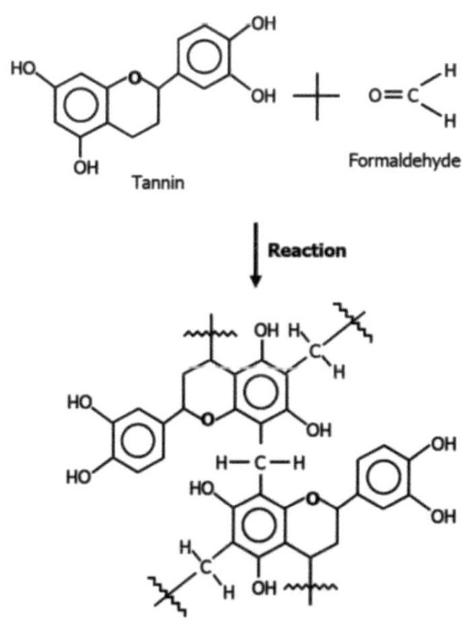

Figure 4.
Reaction mechanism of tannin with formaldehyde.

Formaldehyde is a major aldehyde used for the synthesis, setting, and curing of tannin resin adhesives. It is normally used as a liquid formalin solution or in the form of the polymer paraformaldehyde, which is capable of fairly rapid depolymerization under alkaline conditions. The formaldehyde reaction with tannin can be controlled by the addition of alcohols to the system. Under these circumstances, some of the formaldehydes are stabilized by the formation of hemiacetals, such as the formation $CH_2[OH][OCH_3]$, if methanol is used. When the adhesive is cured at an elevated temperature, the alcohol is driven off and formaldehyde is progressively released from the hemiacetal. These effects minimize formaldehyde volatilization when the reactants reach curing temperature and extend the pot life of the adhesive.

Hexamethylenetetramine (hexamine) may also be added to tannin resins owing to its formaldehyde-releasing action under heat. Although hexamine is unstable in acidic environments, formaldehyde is liberated under alkaline conditions when heated. This effect indefinitely extends pot life at the room temperature. However, in most cases, hexamine does not decompose formaldehyde and ammonia in the presence of chemical species with highly reactive nucleophilic sites, such as melamine, resorcinol, and condensed flavonoid tannins. Instead, unstable intermediate fragments can be reacted with highly reactive nucleophilic sites, such as tannin or melamine, among others, to form amino methylene bridges before yielding formaldehyde. Any species with a strong negative charge under alkaline conditions can react with the intermediate species formed by the decomposition of hexamine far more readily than formaldehyde. This characteristic accounts for the capability of wood adhesive formulations based on hexamine to render bonded panels with extremely low formaldehyde emission [13].

In the absence of highly reactive species with strong negative charges, hexamine decomposition proceeds rapidly and results in formaldehyde formation. Formaldehyde emissions from wood particleboards bonded with pine and wattle tannin-based adhesives with paraformaldehyde, hexamine, and tris(hydroxyl) nitromethane hardeners have been measured using the perforator method. All particleboards manufactured using wattle tannin systems with three different hardeners satisfied grade E1 requirements. On the contrary, only particleboards made with pine tannin and hexamine hardener satisfied grade E1 requirements. This tendency was attributed to the curing mechanism of the hardener, the reactivity of the tannin molecule toward formaldehyde, and rapid reactivity of pine tannin toward formaldehyde [13, 14].

Formaldehyde is substituted with other aldehydes given that the methylene linkages may be too short to form cross-linkages. Pizzi and Scharfetter have shown that furfural-aldehyde is an efficient cross-linking agent and an excellent plasticizer for tannin resin adhesives [15, 16]. The complete replacement of formaldehyde with other aldehydes is unfeasible owing to their slow reactivity with tannins. For example, the water resistance of cured tannin-formaldehyde networks was improved by substituting 10–30% of formaldehyde with other aldehydes with saturated hydrocarbon chains but not by the cosmetic addition of water repellents such as waxes. Tannin adhesives prepared and/or set and/or cured with other adhesives only or with mixtures of formaldehyde and high proportions of other aldehydes yielded cured bonds weaker than those obtained with formaldehyde alone or its mixtures with furfural.

The metal ion effect on phenol-formaldehyde reactions can be applied to condensed tannins of the flavonoid type with some degree of success. The acceleration effect of the metal ions follows the order of Pb^{II}, Zn^{II}, Cd^{II}, $Ni^{II} > Mn^{II}$, Mg^{II}, Cu^{II}, $Co^{II} > Mn^{III}$, $Fe^{III} \gg Be^{II}$, $Al^{III} > Cr^{III}$, Co^{III}.

4.2 Acidic and alkaline hydrolysis and autocondensation

Tannin is subjected to two competing reactions when heated in the presence of strong mineral acids: (1) degradation leading to anthocyanidin and catechin formation and (2) condensation as a result of the hydrolysis of heterocyclic rings (p-hydroxybenzyl ether links). The created p-hydroxybenzyl carbonium ions condense randomly with nucleophilic centers on other tannin units to form phlobaphenes. Other modes of condensation such as free radical coupling of B-ring catechol units cannot be excluded in the presence of atmospheric oxygen [17].

The interflavonoid bonds of condensed tannins with phloroglucinolic A-rings are susceptible to cleavage under even mild alkaline conditions. This characteristic could increase the reactivity with aldehydes. Increased reactivity and autocondensation can be introduced through heterocyclic ring opening.

A drastic increase in the reactivity can be attributed to the liberation of the phloroglucinol species of intermediate products. Model compounds have been used to demonstrate that alkaline-catalyzed rearrangements increase tannin reactivity. Nevertheless, some researches have considered model compounds to demonstrate that tannin structural rearrangements can increase or decrease reactivity toward aldehydes.

The autocondensation reactions that are characteristic of polyflavonoid tannins have recently been utilized in adhesive preparation processes, i.e., adhesive hardening in the absence of aldehyde. Autocondensation reactions are based on the opening of the O1–C2 bond of the flavonoid repeat unit and the subsequent condensation of the reactive center formed at C2 with free C6 or C8 sites of a flavonoid unit on another tannin chain under alkaline or acidic conditions (**Figure 5**). Although this reaction increases the viscosity considerably, gelling does not generally take place. Normally, gelling occurs (1) in the presence of a small amount of dissolved silica (silicic acid or silicates) catalyst or some other catalysts and (2) on a lignocellulosic surface.

In the case of highly reactive pine tannin, cellulose catalysis is sufficient to induce hardening and to produce boards with strengths that satisfy the relevant standard requirements for interior-grade panels. The addition of dissolved silica or silicate catalyst to low-reactive tannins, such as mimosa and quebracho, is the best approach to achieve the required panel strength. The amount of silicic acid or silicates affects gelling. Gelling accelerates as silicate content increases and stabilizes after reaching a certain value. Although tannin resin adhesive that was manufactured through autocondensation increases the dry strength of panels, the strength of the resulting crosslinking is insufficient for exterior-graded panels [18]. Aldehyde curing agents should be added for the preparation of exterior-graded panels. Nevertheless, hardening through tannin autocondensation without any aldehyde addition is also possible. The mechanism of polyflavonoid autocondensation has been examined using carbon-13 nuclear magnetic resonance and electron-spin resonance spectroscopy, among others [19–21].

Zinc acetate also appears to induce a similar type of autocondensation reaction that is slower than that induced by an aldehyde. The reaction induced by zinc acetate mainly occurs at high curing temperatures. Consequently, the effect of zinc acetate is too weak to hinder interflavonoid bond cleavage and pyran ring opening in procyanidins. Therefore, in the presence of zinc acetate, the autocondensation of prodelphinidins to prodelphinidins and prodelphinidins to procyanidins will occur, whereas that of procyanidins to procyanidins will never or will rarely occur [22].

The autocondensation of polyflavonoid tannin is facilitated by the reaction that occurs on cellulose and lignocellulosic substrates. Cellulose-induced polyflavonoid

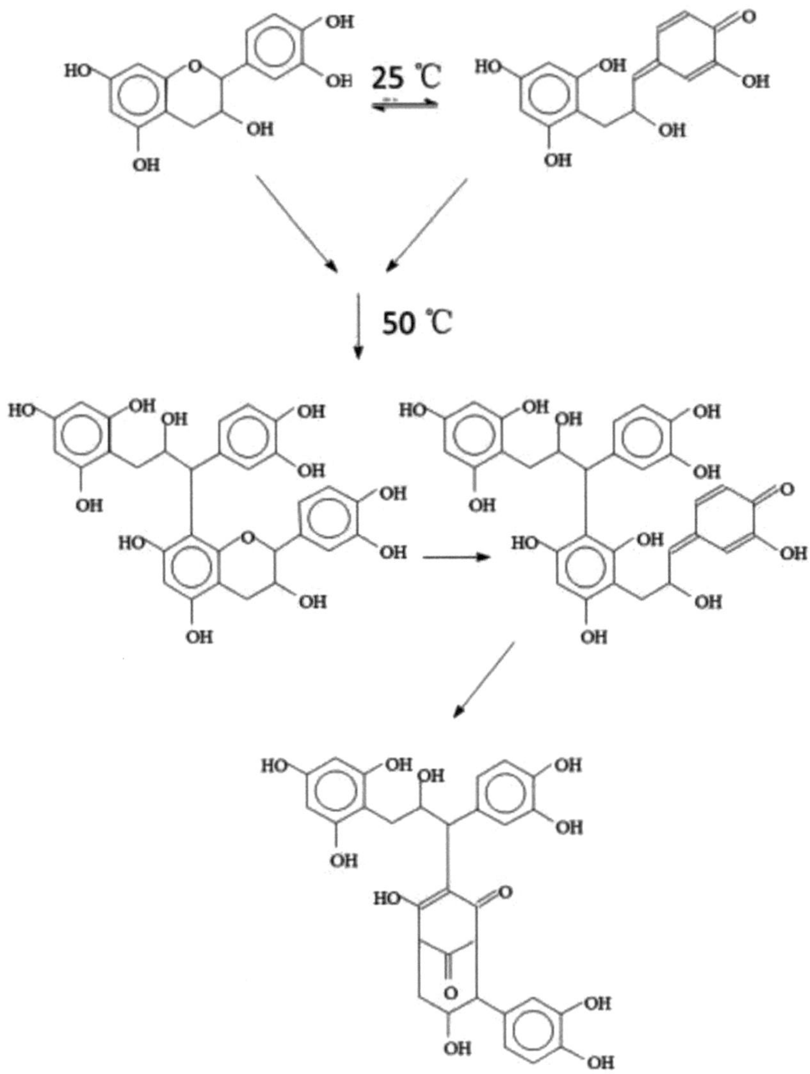

Figure 5.
Autocondensation of tannin resin.

autocondensation and Lewis acid-induced polyflavonoid autocondensation have different mechanisms but involve similar subsequent reactions [23].

4.3 Sulfite reaction

Tannin sulfonation is one of the most useful reactions in flavonoid chemistry and can be particularly useful for the preparation of tannin resin adhesives. The drastic differences between the sulfite treatment products of resorcinol A-ring type tannins (e.g., black wattle tannins) and those of resorcinol A-ring type tannins (e.g., pine tannins) are mainly attributed to the different stabilities of the linkage bonds between their units relative to those of heterocyclic ether bonds. When sodium bisulfite is used to treat black wattle tannins, heterocyclic ether bonds first open because of the relative stability of the connecting bonds between units. Then, sulfonate is added to C-2. In this situation, tannin molecules are negligibly degraded.

The reaction of 5,7-dihydroxy proanthocyanidins with sulfite ions under normal pH conditions proceeds through the cleavage of the interflavonoid bond with the formation of flavan-4- or proanthocyanidin-4-sulfonates, as indicated by the scheme shown in **Figure 6**.

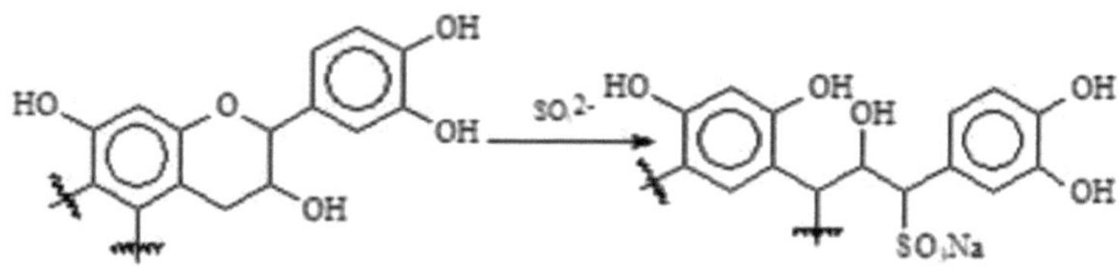

Figure 6.
Sulfonation of tannins.

Sulfonated products can be obtained from phloroglucinolic tannins without the opening of the etherocyclic ring because interflavonoid bonds are easily cleaved. Flavan-2,4-disulfonates are also formed readily.

The involvement of interflavonoid bond cleavage in the sulfonation of phloroglucinolic condensed tannins affects the utilization of these tannins because their molecular weights can be tailored to suit their applications such as wood adhesives. Additionally, sulfonation affords tannins with reduced viscosity and increased solubility through the following mechanisms:

1. The elimination of the water-repellent etherocyclic ether group.

2. The introduction of the hydrophilic sulfonate group and another hydroxyl group.

3. The reduction in polymer rigidity, steric hindrance, and intermolecular hydrogen bonding through the opening of the etherocyclic ring.

4. The hydrolysis of hydrocolloid gums and interflavonoid bonds under acidic conditions.

However, sulfonation may be disadvantageous because sulfonate groups promote sensitivity to moisture and thus aggravate the deterioration of adhesive. This problem could be solved through desulfonation. The desulfonation of 2,4,6-trihydroxybenzyl sulfonic acid and sodium epicatechin-(4β)-sulfonate is a facile reaction under mild alkaline conditions (i.e., pH > 8.0 and ambient temperature). Hydroxyl benzyl sulfonic acids with resorcinol or phenol functionalities resist desulfonation at a pH value of 12 and a temperature of 90°C. Therefore, sulfonation not only reduces molecular weight while improving the viscosity and solubility of tannin resin adhesives but also prevents sulfonic acid functionalization and affords aldehyde condensation products that are insoluble in water [24].

5. Applications

Tannin resin adhesives can be cured under high heat (thermosetting) or at room temperature (coldsetting) [25]. Thermoset tannin resin adhesives are used in the preparation of plywood, particleboard, wood preservation resin, and impregnated resin, among other wood composites. Coldset tannin resin adhesives are used to manufacture glulam, laminating veneer lumber, and finger joints.

5.1 Application of thermoset tannin resin adhesives in glued wood products

5.1.1 Plywood

Tannin resin adhesives are used to prepare plywood (**Figure 7**). However, adhesives manufactured with conventional formulations and technology fails to meet the requirements set for the exterior plywood adhesive. Thus, tannin resin adhesives must be modified by mixing them with other synthetic resins or organic or inorganic modifiers and by optimizing resin synthesis parameters and hot-pressing conditions. Additives can effectively solve the problems of tannin resin hydrophobicity and formaldehyde release and can improve the physical and mechanical properties, especially weather resistance, of the final wood products. Plywood products assembled with modified tannin resin adhesives meet the demand of exterior-grade plywood, have better properties than plywood assembled using phenolic resin, and have a certain commercial potential [26–30].

5.1.2 Particleboard

The use of tannin resin adhesives in particleboard production has been accepted in many countries and is used in the manufacturing of industrial particleboard in many countries except in Asian countries. For example, mimosa tannin resin adhesives are used in industrial particleboard manufacture in South Africa and South America (**Figure 8**).

Although tannin resin adhesives have been successfully used for the production of interior and exterior particleboards, the synthesis processes and formulas of tannin resin are drastically different [31]. Generally, tannin resins polymerized with formaldehyde have higher weathering resistance than those polymerized with other nontoxic and nonvolatile aldehydes in accordance with the reactivity of tannin. Particleboards manufactured using tannin resin adhesives with formaldehyde contents that have been partially or completely replaced with acetaldehyde have very low formaldehyde emission or even free from formaldehyde release and have mechanical properties that meet the requirements of interior-grade application [32]. At the same time, particleboards prepared with tannin resin synthesized via an aldehyde-free process can also meet the requirements of interior-grade applications [33]. An appropriate amount of paraformaldehyde or curing agent must be added to increase the weathering resistance of particleboards prepared with tannin resin

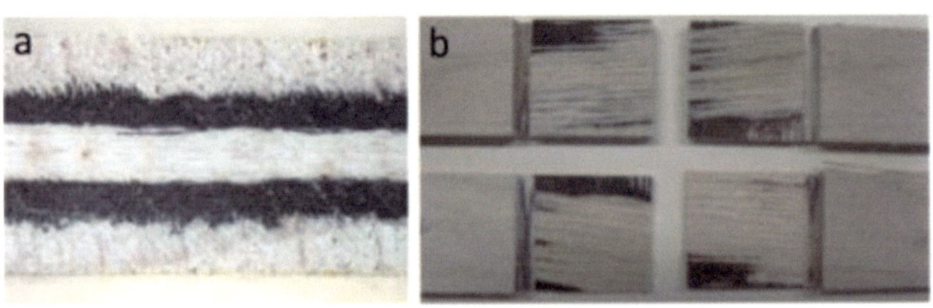

Figure 7.
Tannin resin adhesives for plywood manufacturing: (a) profile image; (b) wood failure.

Figure 8.
Tannin resin adhesives for manufacturing particleboard.

adhesives. Sometimes phenolic resin is also mixed with tannin resin. The properties of particleboards produced with 60% tannin resin still meet the requirements of exterior-grade application [34].

Tannin resin adhesives for particleboard production have high requirements for curing agents. Different types of tannin resin adhesives require different curing agents. The performances of tannin, i.e., tannin structure and curing agent selection, and the properties of the particleboard will be affected by hot-pressing conditions, including pressing temperature, time, and pressure. Selecting an appropriate curing agent can accelerate tannin resin curing, reduce formaldehyde emission, and most importantly, can ensure that the performances of the particleboard meet exterior-grade application requirements [35]. Kim et al. found that the reactive speeds of tannin resin adhesives for black wattle tannin followed the order of paraformaldehyde > hexamine > trinitromethane, whereas those for pine tannin followed the order of hexamine > paraformaldehyde > trinitromethane [36].

Additionally, modified tannin resin can be used to prepare particleboards from different sources, such as wheat straw [37], rice husk [38], cashew nut shell [39], and chestnut shell [40]. The elastic moduli, internal bonding strength, and water-absorbing thickness swelling of the prepared particleboards meet the requirements of European standards.

5.1.3 Wood preservation

Wood preservation is vital for protecting wooden products. Traditional approaches for wood preservation include the treatment of wood with various chemical agents to prevent attack by different organic microorganisms and insects. Although traditional wood preservatives confer good effects and strong durability, they inevitably introduce various other problems, such as environmental pollution and carcinogenic effects. Tannin is a natural fungicide and good preservative that can be used to prevent fungal and bacterial damage [41]. Most plant pathogens secrete enzymes that degrade cellulose and lignin. Tannin can effectively inhibit the activity of these enzymes and prevent the proliferation of pathogens by complexing with protein [42]. Pizzi and Conradie confirmed that the antifungal activity of wood treated with flavonoid tannin resins is twice as intense as that of neat wood [43]. Additionally, veneer treated with tannin resin modified with a small amount of boric acid avoided the fungal attack and exhibited high durability, mechanical strength, and fire resistance because tannin and boric acid can simultaneously inhibit bacterial and fungal growth (**Figure 9**). Meanwhile, the fixation of boric acid in wood with tannin resin and hexamine prevented loss and exerted a good preservative effect that met the requirements of the European standard EN 113 [44–46].

Figure 9.
Tannin resin adhesives for wood preservation.

5.1.4 Wood-impregnated paper

Melamine formaldehyde resin has been widely used in the decorative impregnated paper industry. The addition of small amounts of urea can drastically cut costs without affecting performance. Melamine resin-impregnated paper is widely used for the production of laminated wooden floors and panel furniture overlays [47]. Phenolic resin-impregnated paper has limited applications in the production of floor and furniture panels owing to its black color, but it is widely accepted and popular in some particular applications, such as the impregnation of architectural template veneer paper, owing to its good adhesive property and high strength [48].

The flavonoid phenol structure and properties of tannin are similar to those of phenols. Therefore, there are no theoretical constraints for using tannin resin in paper impregnation. Abdullah et al. [49, 50] synthesized a low-viscosity tannin resin, which was used to impregnate paper with a glue amount of 172 g/m^2 and hot-pressed on wood-based panel substrates. The final overlay paper exhibited a very smooth surface, high wear resistance, scratch resistance, and water steam resistance. The performances of the optimized overlay paper were even superior to those of overlay paper impregnated with MUF resin (**Figure 10**). Similar to phenolic resins, overlay paper impregnated with tannin resin has potential use in template production.

5.1.5 Fiber: veneer composites

Natural fiber composites have been developed to overcome the limitations associated with petrochemical resources. They have extensive prospective applications in the automobile and aerospace industries because their raw materials are derived from biomass and they possess unique characteristics. Fibers impregnated with

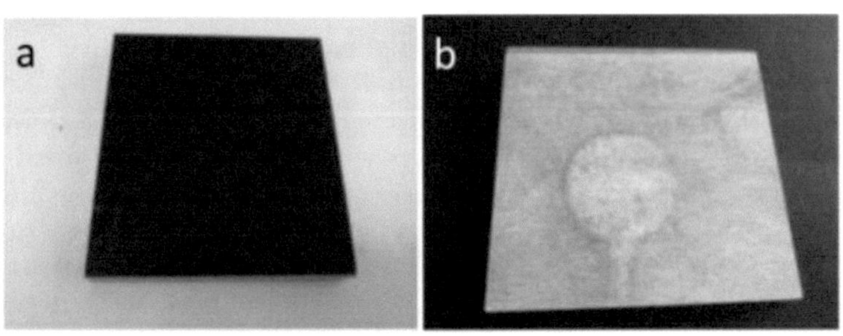

Figure 10.
Overlay paper impregnated with (a) tannin resin; (b) MUF resin.

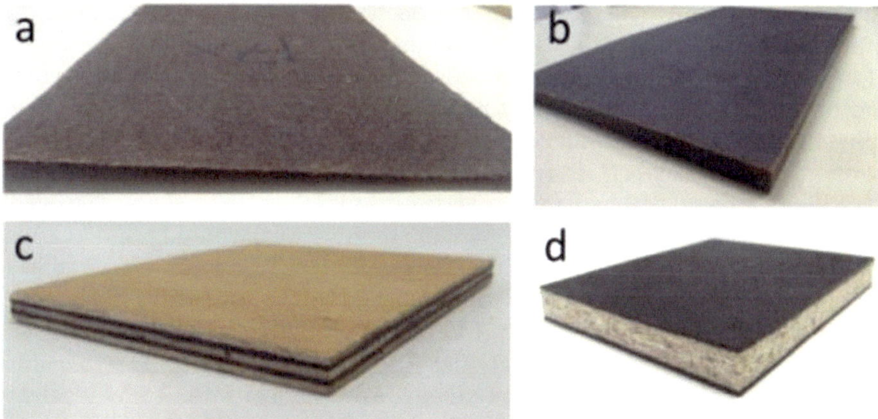

Figure 11.
(a) Tannin resin impregnated fiber; (b) laminated composite with tannin resin impregnated fiber; (c) wood composites with veneer and tannin resin impregnated fiber; (d) tannin resin impregnated fiber overlay on the particleboard.

tannin resin can be used to synthesize different types of fibers or wood-based fiber composites (**Figure 11**). These composites possess high elastic moduli and tensile strength and good water-absorbing expansibility [51–54].

5.1.6 Other wooden panels

In addition to wooden panels, medium density fiberboard [55], oriented strand boards [56], wafer boards [57], container boards [58], and other furniture panels [59] could be prepared with tannin resin adhesives.

5.2 Application of coldset tannin resin adhesives in glued wood products

Adhesives for finger joint lumber and glulam must meet high standard requirements because of the rigorous application environment of these materials (**Figure 12**). These adhesives must possess high mechanical strength and

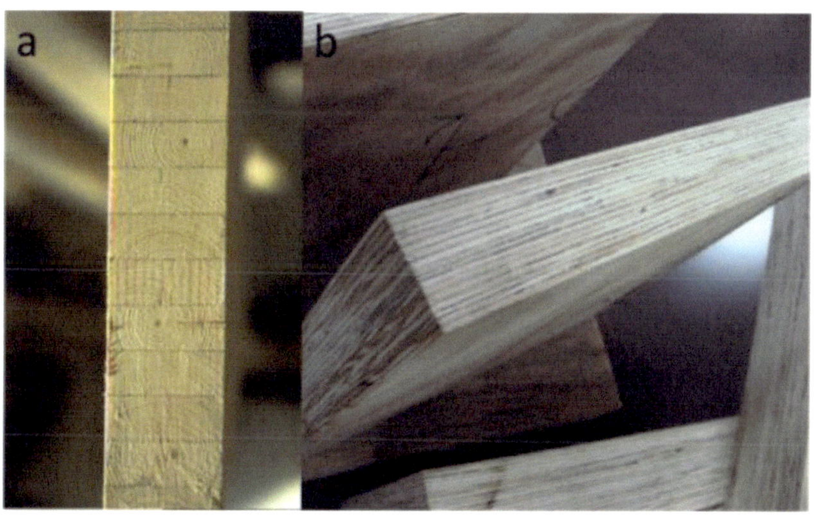

Figure 12.
Coldset tannin resin adhesives for wooden product application: (a) glulam; (b) laminates.

weather resistance. Acacia tree tannin has been widely used to prepare low-temperature curing resin adhesives for finger joint and laminated veneer lumbers [60]. Acacia tannin resin adhesive has a low curing temperature and excellent bonding performance. Additionally, the cost of acacia tannin is lower than that of phenol resorcinol formaldehyde resin. The performance and cost of tannin-resorcinol-formaldehyde resin adhesives must be balanced. The tannin content of such adhesives could reach up to 95% after optimization. The polymerization of resorcinol units is replaced by that of a large number of flavonoid tannin natural phenol units. Resin adhesives with high tannin contents can be cured at room temperature, can exhibit good performances, and can be used to produce veneer and finger joint lumbers and glulam [61, 62]. Other nonvolatile or nontoxic aldehydes, such as glyoxal and glutaraldehyde, are also used to synthesize coldset tannin resin adhesives to effectively reduce the problem of formaldehyde release. Although only a small amount of paraformaldehyde is used to cure tannin resin adhesives, the prepared wooden products demonstrate good mechanical strength and water resistance [63].

6. Conclusions

Tannin has significant application prospects as a promising natural phenolic polymer. However, this raw material continues to exhibit limitations, such as reactivity, high viscosity, short pot life, and poor weather resistance, among others. Future works must address these problems. Research on tannin resin adhesives should focus not only on wood panels but also on other advanced wooden composites. In addition, the industrialization of tannin resin adhesives in the field of wood manufacturing field is necessary.

Acknowledgements

This work is supported by the following grants and programs: 1. National Natural Science Foundation of China (NSFC 31760187); 2. Yunnan Provincial Applied and Basic Research Grants (2017FB060); 3. "Ten-thousand Program"-youth talent support program; and 4. Yunnan Provincial Reserve Talents for Middle & Young Academic and Technical Leaders.

Conflict of interest

There is no conflict of interest in this field.

Author details

Xiaojian Zhou* and Guanben Du
Yunnan Provincial Key Laboratory of Wood Adhesives and Glued Products, Southwest Forestry University, Kunming, China

*Address all correspondence to: xiaojianzhou@hotmail.com

References

[1] Dunky M, Pizzl A. Chapter 23–Wood adhesives. In: Adhesion Science & Engineering. 2002. pp. 1039-1103

[2] Tondi G. Tannin-based copolymer resins: Synthesis and characterization by solid state 13C NMR and FT-IR spectroscopy. Polymers. 2017;**9**(6):223

[3] Roffael E, Dix B, Okum J. Use of spruce tannin as a binder in particleboards and medium density fiberboards (MDF). European Journal of Wood and Wood Products. 2000;**58**(5):301-305

[4] Drewes E, Roux DG. Condensed tannins. XV. Interrelations of flavonoid components in wattle-bark extract. The Biochemical Journal. 1963;**87**(1):167-172

[5] Roux DG, Paulus E. Condensed tannins. VIII. The isolation and distribution of interrelated heartwood components of Schinopsis species. The Biochemical Journal. 1961;**78**:785-789

[6] Pasch H, Pizzi A, Rode K. MALDI-TOF mass spectrometry of polyflavonoid tannins. Polymer. 2001;**42**:7531-7539

[7] Pizzi A. Mechanism of viscosity variations during treatment of wattle tannins with hot NaOH. International Journal of Adhesion and Adhesives. 1998;**4**(1):213-214

[8] Sowunmi S, Ebewele RO, Conner AH, et al. Differential scanning calorimetry of hydrolysed mangrove tannin. Polymer International. 2000;**49**:574-578

[9] Pizzi A, Horak RM, Roux DG. Condensates of phenol, resorcinol, phloroglucinol, and pyrogallol as model compounds of flavonoid A- and B-rings with formaldehyde. Journal of Applied Polymer Science. 1979;**24**:1571-1578

[10] Pizzi A. Hot-setting tannin-urea-formaldehyde exterior wood adhesives. Adhesives Age. 1977;**12**:27-29

[11] Pizzi A, Merlin M. A new class of tannin adhesives for exterior particleboard. International Journal of Adhesion and Adhesives. 1981;**1**(5):261-264

[12] Pizzi A. Exterior wood adhesives by MDI crosslinking of polyflavonoid tannin B rings. Journal of Applied Polymer Science. 1980;**25**:2123-2127

[13] Pizzi A, Tekely P. Mechanism of polyphenolic tannin resin hardening by hexamethylenetetramine: CP-MAS 13C-NMR. Journal of Applied Polymer Science. 1995;**56**:1645-1650

[14] Kim S, Kim HJ. Evaluation of formaldehyde emission of pine and wattle tannin-based adhesives by gas chromatography. Holz als Roh- und Werkstoff. 2004;**62**:101-106

[15] Pizzi A, Scharfetter HO. The chemistry and development of tannin-based adhesives for exterior plywood. Journal of Applied Polymer Science. 1978;**22**:1745-1761

[16] Baaka N, Ammar M, Saad MK, et al. Properties of tannin-glyoxal resins prepared from lyophilized and condensed tannin. Journal of Fashion Technology & Textile Engineering. 2017;**3**(4):00110

[17] Sun WD. Chemistry of Tannin. China: China Forestry Publishing House; 1988. pp. 368-388

[18] Pizzi A, Meikleham N. Induced accelerated autocondensation of polyflavonoid tannins for phenolic polycondensates. II. CP-MAS 13C-NMR of different tannins and models. Journal of Applied Polymer Science. 1995;**55**:1265-1269

[19] Garcia R, Pizzi A. Polycondensation and autocondensation networks in polyflavonoid tannins. I. Final networks. Journal of Applied Polymer Science. 1998;**70**:1083-1091

[20] Masson E, Pizzi A, Merlin M. Comparative kinetics of the induced radical autocondensation of polyflavonoid tannins. II. Flavonoid units effects. Journal of Applied Polymer Science. 1997;**64**:243-265

[21] Garcia R, Pizzi A, Merlin A. Ionic polycondensation effects on the radical autocondensation of polyflavonoid tannins: An ESR study. Journal of Applied Polymer Science. 1997;**65**:2623-2633

[22] Pizzi A, Tekely P. 13C-NMR of Zn2+ acetate-induced autocondensation of polyflavonoid tannins for phenolic polycondensates. Journal of Applied Polymer Science. 1995;**56**:633-636

[23] Pizzi A, Meikleham N, Stephanou A. Induced accelerated autocondensation of polyflavonoid tannins for phenolic polycondensates. II. Cellulose effect and application. Journal of Applied Polymer Science. 1995;**55**:929-933

[24] Lei H. Résines à base de matériaux naturels et synthétiques destinées aux adhésifs pour le bois. In: Nancy 1. 2009

[25] Brosse N, Pizzi A. Tannins for wood adhesives, foams and composites: Preparation, Characterization, and Testing. In book: Bio-based wood Adhesives. 2017. pp. 197-220

[26] Efhamisisi D, Thevenon MF, Hamzeh Y, et al. Induced tannin adhesive by boric acid addition and its effect on bonding quality and biological performance of poplar plywood. ACS Sustainable Chemistry & Engineering. 2016;**4**(5):2734-2740

[27] Zhao L, Cao B, Wang F, Yazaki Y. Chinese wattle tannin adhesives suitable for producing exterior grade plywood in China. Holz als Roh- und Werkstoff. 1994;**52**(2):113-118

[28] Moubarik A, Pizzi A, Allal A, Charrier F, Charrier B. Cornstarch and tannin in phenol–formaldehyde resins for plywood production. Industrial Crops and Products. 2009;**30**(2):188-193

[29] Pizzi A, Knauff C, Sorfa P. Effect of acrylic emulsions on tannin-based adhesives for exterior plywood. Holz als Roh- und Werkstoff. 1981;**39**(6):223-226

[30] Gao Z, Yuan J, Wang X. Phenolated larch-bark formaldehyde adhesive with multiple additions of sodium hydroxide. Pigment & Resin Technology. 2007;**36**(5):279-285

[31] Pizzi A. Tannin-based adhesives. Journal of Macromolecular Science—Reviews in Macromolecular Chemistry. 1980;**18**(2):247-315

[32] Ballerini A, Despres A, Pizzi A. Non-toxic, zero emission tannin-glyoxal adhesives for wood panels. Holz als Roh- und Werkstoff. 2005;**63**(6):477-478

[33] Moubarik A, Mansouri HR, Pizzi A, Charrier F, Allal A, Charrier B. Corn flour-mimosa tannin-based adhesives without formaldehyde for interior particleboard production. Wood Science and Technology. 2013;**47**(4):675-683

[34] Lu Y, Shi Q. Larch tannin adhesive for particleboard. Holz als Roh- und Werkstoff. 1995;**53**(1):17-19

[35] Pizzi A, Valenezuela J, Westermeyer C. Low formaldehyde emission, fast pressing, pine and pecan tannin adhesives for exterior particleboard. Holz als Roh- und Werkstoff. 1994;**52**(5):311-315

[36] Kim S, Lee Y, Kim H, Lee HH. Physico-mechanical properties of particleboards bonded with pine and wattle tannin-based adhesives. Journal

of Adhesion Science and Technology. 2003;**17**(14):1863-1875

[37] Tabarsa T, Jahanshahi S, Ashori A. Mechanical and physical properties of wheat straw boards bonded with a tannin modified phenol–formaldehyde adhesive. Composites Part B: Engineering. 2011;**42**(2):176-180

[38] Tesha JV, Ndazi B, Bisanda E. Properties of rice husk particleboards bonded by a blend of tannin-cnsl resin. Tanzania Journal of Engineering and Technology. 2018;**25**(2):55-64

[39] Bisanda E, Ogola W, Tesha J. Characterisation of tannin resin blends for particle board applications. Cement and Concrete Composites. 2003;**25**(6):593-598

[40] Santos J, Antorrena G, Freire MS, et al. Environmentally friendly wood adhesives based on chestnut (*Castanea sativa*) shell tannins. European Journal of Wood and Wood Products. 2017;**75**(1):89-100

[41] Zucker WV. Tannins: Does structure determine function? An ecological perspective. The American Naturalist. 1983;**121**(3):335-365

[42] Laks PE, McKaig PA, Hemingway RW. Flavonoid biocides: Wood preservatives based on condensed tannins. Holzforschung-International Journal of the Biology, Chemistry, Physics and Technology of Wood. 1988;**42**(5):299-306

[43] Pizzi A, Conradie W. A chemical balance/microdistribution theory-new CCA formulations for soft-rot control? Material und Organismen. 1986;**21**(1):31-46

[44] Moubarik A, Charrier B, Charrier F, Pizzi A, Allal A. Evaluation of decay resistance of wood products made from borax-impregnated wood and bonded with a formaldehyde-free cornstarch and tannin adhesive. Annals of Forest Science. 2009;**66**(1):1

[45] Thevenon M, Tondi G, Pizzi A. High performance tannin resin-boron wood preservatives for outdoor end-uses. European Journal of Wood and Wood Products. 2009;**67**(1):89-93

[46] Thévenon M, Tondi G, Pizzi A. Friendly wood preservative system based on polymerized tannin resin-boric acid for outdoor application. Maderas-Ciencia Y Tecnologia. 2010;**12**(3):253-257

[47] Taylor AR, Wulfekotter D. Modified melamine resin for use in decorative laminates: U.S. Patent 4,713,299. December 15, 1987

[48] Chun-gui DU. A study on top grade bamboo-wood composite concrete-form. Wood Processing Machinery. 2004;**15**(1):9-11

[49] Abdullah U, Zhou X, Pizzi A, Merlin A. Note on surface quality of plywood overlaid with mimosa (*Acacia mearnsii*) tannin and melamine urea formaldehyde impregnated paper: Effects of moisture content of resin impregnated papers before pressing on physical properties of overlaid panels. International Wood Products Journal. 2013;**4**(4):253-256

[50] Abdullah UH, Pizzi A, Rode K, Delmotte L, Zhou X, Mansouri H. Mimosa tannin resins for impregnated paper overlays. European Journal of Wood and Wood Products. 2013;**71**(2):153-162

[51] Nicollin A, Li X, Girods P, Pizzi A, Rogaume Y. Fast pressing composite using tannin-furfuryl alcohol resin and vegetal fibers reinforcement. Journal of Renewable Materials. 2013;**1**(4):311-316

[52] Sauget A, Nicollin A, Pizzi A. Fabrication and mechanical analysis of mimosa tannin and commercial

flax fibers biocomposites. Journal of Adhesion Science and Technology. 2013;**27**(20):2204-2218

[53] Zhu J, Zhu H, Immonen K, Brighton J, Abhyankar H. Improving mechanical properties of novel flax/tannin composites through different chemical treatments. Industrial Crops and Products. 2015;**67**:346-354

[54] Zhou X, Segovia C, Abdullah UH, et al. A novel fiber–veneer-laminated composite based on tannin resin. The Journal of Adhesion. 2017;**93**(6):461-467

[55] Xiao-dan H. The development on the black wattle tannin adhesive for exterior medium density fiberboard. China National Forest Products Industry. 2003;**30**(5):31-33

[56] Pichelin F, Pizzi A, Frühwald A. OSB adhesives rate of strength development on single strands couples. European Journal of Wood and Wood Products. 2000;**58**(3):182-183

[57] Li G, Lei H, Shuang-bao Z, De-rong Z, Qiu-hui Z. Study on tannin resin *Pinus tabulaeformis* waferboard. Wood Processing Machinery. 2002;**13**(5):20-23

[58] Saayman HM, Brown CH. Wattle-base tannin-starch adhesives for corrugated containers. Forest Products Journal. 1977;**27**(4):21-25

[59] Lee J, Jeon J, Kim S. Green adhesives using tannin and Cashew nut shell liquid for environment-friendly furniture materials. Journal of Korean Furniture Society. 2011;**22**(3):219-229

[60] Pizzi A, Rossouw DT, Knuffel W, Singmin M. 'Honeymoon' phenolic and tannin-based fast-setting adhesive systems for exterior grade finger joints. Holzforschung und Holzverwertung. 1980;**32**(6):140-150

[61] Gornik D, Hemingway R, Tišler V. Tannin-based cold-setting adhesives for face lamination of wood. European Journal of Wood and Wood Products. 2000;**58**(1):23-30

[62] Mansouri HR, Pizzi A, Fredon E. Honeymoon fast-set adhesives for glulam and fingerjoints of higher natural materials content. European Journal of Wood and Wood Products. 2009;**67**(2):207-210

[63] Zhou X, Pizzi A. Tannin–resorcinol–aldehyde cold-set wood adhesives with only formaldehyde as hardener. European Journal of Wood and Wood Products. 2013;**71**(4):537-538

Chapter 7

Applications of Tannins in Industry

Akhlash P. Singh and Sunil Kumar

Abstract

Tannins are water-soluble natural polyphenols mainly present in plant-based materials, including food. Tannins play a very significant role as a raw material for sustainable green industries. Therefore, they are mainly used in diverse types of industries such as leather, feed, fisheries, beverages, etc. They also find application as potential medicinal agents, antioxidants, metal chelators; and cater as inhibitors of harmful pro-oxidative enzymes and of lipid peroxidation process. Recently, several important properties like antiseptics, anticarcinogenic, and anti-inflammatory of tannins have been documented in the human that make them suitable candidates for pharmaceuticals and nutraceutical industries. Because of current concerns related to synthetic compounds used in the human health and food industries, which leave highly adverse effects on the human body and environment, tannins can offer an alternative to these harmful chemicals in recently emerging industries.

Keywords: tannins, nutraceutical, wood, leather, pharmaceuticals industries and antibacterial activity

1. Introduction

Biochemically, tannins are sort of secondary metabolites predominantly available in the plant-based foods and beverages. The name "tannin" is originated from the industrial process of "leather tanning," in which animal hides are converted into leather through downstream processing. It is worthwhile to mention that tannins were used in this process from historic times. On the basis of their presence, various parts of plants such as bark, wood, leaves, seeds, roots, and even the plant galls are the major sources of tannin extractions used for various purposes (**Table 1**). Algae is also rich source of tannin-based compounds such as phlorotannins which comprise of antioxidant, antidiabetic, anti-inflammatory, and antitumor properties evaluated in the case of human [1, 2]. In addition to health benefits, phlorotannins isolated from brown seaweeds have been used in the cosmetics also [3]. Some researchers have defined tannins as "Any phenolic compound of sufficiently high molecular weight containing sufficient hydroxyls and other suitable groups (i.e., carboxyls) to form effectively strong complexes with protein and other macromolecules under the particular environmental conditions being studied" [4].

Plant tannins are a large group of natural phenolic compounds which contain a range of molecular weight between 500 and 3000 Da. Currently, they have been divided into three main subgroups: (1) hydrolysable tannins, (2) condensed tannins, and (3) phlorotannins. Hydrolysable tannins are highly soluble in water; biochemically, they consist of a central core of a carbohydrate (D-glucose) with

IntechOpen

S. no.	Name of plant species	Plant parts	Major components	Medicinal use
1.	*Krameria triandra* L.	Root	Tannic acid, rhataniatannic acid, peculiar acid principle, krameric acid, phlobaphene, phloroglucin, and proanthocyanidins	Chronic diarrhea, menorrhagia, urinary diseases, bleeding from the bowels, bad throat, and antibacterial agents for the eyes, nose, and gums
2.	*Potentilla erecta* (L.) Rauschal	Roots	Pentadigalloylglucose, pedunculagin, epigallocatechin, catechins, and proanthocyanidins	Inflammations, wound healings, diarrhea, inflammation of bowel bacterial, fungal, and viral infections
3.	*Sanguisorba officinalis* L.	Root	Sanguiin H-6	Dysentery and insect bites
4.	*Potentilla kleiniana*	Aerial parts	Agrimoniin and potentillin	Diarrhea, cough, lymphadenitis, and hepatitis
5.	*Syzygium cumini*	Bark	Corilagin and related ellagitannins	Bad throat, asthma, dysentery, and ulcers
6.	*Quercus robur* L.	Bark	Grandinin, castalagin, and glucogallin	Diarrhea, itching, and burning
7.	*Phyllanthus muellerianus* (Kuntze) Exell	Leaves, stem, and bark	Geraniin, phenazine derivative of geraniin, Corilagin, and furosin.	Wound healing
8.	*Geranium thunbergii* Siebold exLindl. & Paxt.	Leaves	Geraniin	Intestinal disorders
9.	*Mouriri pusa* Gardn. (Melastomataceae)	Leaves	Catechins and other condensed tannins	Gastritis and ulcers
10.	*Acacia nilotica* (L.) Willd. exDelile.	Pod	Gallocatechin-gallate, methyl gallate, catechin, catechin gallate, galloylglucose, and epicatechin	Fever, diabetes, and gum diseases
11.	*Diospyros kaki* Thunb.	Fruit	Proanthocyanidin oligomers based on catechin, gallocatechin, catechin-3-O-gallate, and gallocatechin-3-O-gallate	Antiseptic and cardiovascular diseases
12.	*Quercus infectoria* Oliv.	Gall	Tannic acid	Bacterial, fungal, and viral infection

Note: Table indicates that tannin and its components are present in most of the parts of the plants which offered great level of medicinal sources or pharmaceutical agents [59].

Table 1.
Plant species containing tannins and their medicinal use.

its hydroxyl groups or polyol esterified with phenolic compounds such as gallic acid (3,4,5-trihydroxybenzoic acid) or hexahydroxydiphenic acid, which also known as ellagic acid (ellagitannin). Hydrolysable tannins mainly originated from Pentagalloylglucose (2-O-digalloyl-1, 3, 4, 6-tetra-O-galoyl-α-D-glucopyranose), which is a basic structural unit of hydrolysable tannins. The main source of structural diversity among the hydrolysable tannins is the presence of diverse types of oxidative linkages that give rise to oligomeric compounds with molecular weight between 2000 and 5000 Da [5]. Characteristic examples of hydrolysable tannins are (1) gallic acid; (2) hexahydroxydiphenic acid; (3) ellagic acid; and (4) pentagalloylglucose, which contain a central glucose molecule as the core attached with multiple gallic acid units, while ellagitannins are associated to hexahydroxydiphenic acid. Hydrolysable tannins are mainly present in angiosperm and dicotyledons. Both gallotannins and ellagitannins may synthesize individually or in the form of a mixture in plants. Gallic acid has been extracted from plant families, for example, Ericaceae, Geraniaceae, or Fagaceae; whereas, ellagic acid is available in Hamamelidae, Dilleniidae, and Rosidae species [6].

Naturally occurring condensed tannins are polyphenolic bioflavonoids, are polymers of polyhydroxy flavan-3-ol units, for example, (+)-catechin and (−)-epicatechin-2, (+)-gallocatechin, and flavan-3, 4-diols. They are also known as proanthocyanidins (PA) that ascribed to their hydrolysis to anthocyanidins in heated ethanol treatment. Due to presence of stereoisomerisms in hydroxylation patterns at three chiral centers, bond positions, and type of interflavan bond; proanthocyanidins are present in variety of active forms, for example, (1) (+)-catechin; (2) (−)-epicatechin; (3) (+)-gallocatechin; (4) (−)-epigallocatechin; and (5) (−)-epigallocatechin gallate. Among them, (+) catechin and (−) epicatechin are predominantly present in nature [7]. There are several plant species which offer rich source of proanthocyanidins (see **Table 1**). Proanthocyanidins can be obtained from red wine, green tea, cocoa, and chocolate. However, condensed tannins can further be classified on the basis of reaction rate like, slow reacting tannins like quebracho and mimosa and fast reacting tannins like pine and pecan (*Carya illinoensis*) [5, 7, 8]. Furthermore, some researcher also classified tannins into four different classes based on their structural properties, namely, (1) gallotannins, (2) ellagitannins, (3) complex tannins, and (4) condensed tannins.

After the industrial revolution, most of the synthetic chemicals were used in the diverse types of industries including food, pharma, beverage, leather, and other industries. But prolonged applications of synthetic chemicals in the area of health and other industries left a myriad of adverse effects on environment and human health. Therefore, current focus has been shifted on alternative natural compounds like tannins that can be exploited in the form of functional food, nutraceutical, cosmetology, and pharmaceutical industries.

The global tannin market is expanding very rapidly; according to estimation, 1076.3 kilotons tannin was required in 2015 which expected to rise with CAGR of 5.8% from 2016 to 2025. The demand was mainly in the wine, leather, pharmaceuticals, and wood industries. In case of USA, approximately 282.4 kilotons tannins were produced; its 62.3% is used in leather industries alone. Europe is another emerging market for tannins and tannin-based products. It is due to large scale wine production, which is accounted for 38% revenue generations in Europe. Hence, global market size for tannin related industries may increase up to $3.3 billion by 2025. The applications of condensed tannins are expected to increase up to 424.8 kilotons by 2025 in comparison of 242.9 kilotons in 2015 [9].

But, there are several unresolved issues linked with applications of tannins such as antinutrient effect, resistance to enzymatic hydrolysis, and lack of complete information about their interactions with other biomolecules and mode of actions in human and animals. The aim of this chapter is to present a brief discussion on the application of tannins in modern industries and to review their positive and negative aspects. It also shows that tannins are being used as sustainable raw material with other green materials in new emerging industries.

2. Application of tannins in various industry

Currently, hydrocarbon-based raw materials are exploited in different petrochemical industries ranging from fuel to cosmetology. It leads to the widespread deficiency of raw material eventually that creates high inflations, environmental degradation, and adverse effects on human and animal health. This necessitates to explore new alternative natural biopolymers such as polylactic acid, chitosan, lignin, and tannins for replacing with currently used hydrocarbon based polymers. Tannins can be the best natural raw material for emerging and traditional industries. This is attributed to tannin's unique natural properties, chemical structure, and commercial properties [10]. Tannins provide several advantages like being as good biomaterial, antimicrobial, antioxidant, pharmaceutical, biopesticide, and nutraceutical agent. Tannins can be tapped for their applications in food, wood, leather, pharma, and other industries as possible raw material, as given below.

2.1 Food industry

Tannins are the secondary metabolites present in a substantial amount in plant-based food products. Due to their positive effects on the food as antibacterial and antioxidants, they are the major constituent of foods. Tannins are used as food preservatives, packaging materials, and food enhancements which owe to their protective nature.

2.1.1 Food packaging

Currently, most of the food items available in the market are wrapped in the packing materials which are plastic, polyethylene, and low-density polyethylene (LDPE) and linear low-density polyethylene (LLDPE) due to their lightness, inertness, and easy availability. In fact, packaging increases the shelf-life and prevents physical damage, contamination, and deterioration in view of environmental contaminants. But these synthetic materials pose great level of environmental and health hazards. Hence, the concept of natural and active packaging has been introduced and there being continuous efforts to make packing materials from biological sources like, chitosan, starch, gelatin, tannins, and methylcellulose [11, 12]. In view of consumer awareness and knowledge, it becomes essential to develop new wrapping material for food items. However, nitrocellulose-based package are already in use in food industry, but currently the "active packaging" also introduced. Active packaging is a better option which protects the food material, simultaneously it also acts as an additive to improve antioxidant properties of foods and absorb unwanted substances such as heavy-metals or exhausted oils and to protect against oxidation, UV, and moisture-based degradation.

Recently, a packaging material was prepared by introducing tannins into cellulose nanofibrils in a single step process of mechanical fibrillation. This newly developed packaging film offers high density, and enhanced surface hydrophobicity which resulted in almost six times improvement in air-barrier and antioxidants properties. Simultaneously, nanocellulose-tannin-based films are active packaging materials which also provide a green, sustainable, nontoxic packaging source for food and pharmaceutical products [11, 12].

Two perishable food items, Cherry tomatoes (*Solanum lycopersicum var. cerasiforme*) and grapes (*Vitis vinifera*), were effectively preserved from *Escherichia coli* (*E. coli*) and *Staphylococcus aureus* (*S. aureus*) for 14 days by tannin-based film that was produced by introducing 15% (w/w) tannin into chitosan, gelatin, and methylcellulose films. Hence, tannins improve the physiochemical properties of biopolymer films, consequently improvement in overall ability to preserve fruits and vegetables. In addition, it reduced the weight loss and improved the browning index of fruits during storage. Another oligomeric procyanidins (OPCs)-based membrane in combination of flaxseed gum (FG) and lauric acid (LA) had more water vapor permeability (WVP), mechanical properties, and peroxide value (POV) of packaging film [13]. The film was evaluated for oil, salt, and vegetable preservation for 75 days. The results indicate that membrane can be used as promising packaging material.

Generally, chitin is used as packing material, but its poor antibacterial and antioxidant abilities make it unsuitable for food packaging; therefore, tannic acid was introduced in the chitin film via single step process of interfacial assembly. Tannins addition to packaging has significantly improved the antibacterial and antioxidant properties of chitin-based packaging film. Improved quality of chitin-based film is mainly attributed to hydrogen and hydrophobic bond formation between chitin and tannins [14].

Protein is a major component of food, and its protection is of utmost significance in packed foods. So far, tannin- and carbohydrate-based packaging films were prepared. But, a soluble dietary fiber (SDF) and tannin-based nanocluster assembly is prepared by introducing calcium ions that creates a cross-linking nucleus in membrane. Linkage between nanoassembly and proteins offer additional advantages such as high level of antimicrobial properties and excellent cell biocompatibility which was proved by FT-IR, XRD, and DSC tests [15]. Tannin-based packing materials thus offer green, sustainable, and ecofriendly alternative that can be used in the food preservation and biomedical fields.

2.1.2 Food preservation

Microorganism, fungus, yeasts, virus, pollens, and chemicals are the biggest threat to food's shelf-life in home as well as in food markets. Biochemically, proanthocyanidins and gallic acid are flavonoid monomers by nature and are major food constituents isolated from pomegranate, strawberry, blackberry, raspberry, walnuts, almonds, and seeds. Various studies have proved that tannins prevent growth of microorganisms. Tannins are quite effective against the resistant methicillin-resistant *Staphylococcus aureus* (MRSA) [16, 17]. Moreover, Punicalagin, an ellagitannin, isolated form pomegranate peel show very strong antibacterial properties against *Staphylococcus aureus* and can be used to control *S. aureus* contamination in food industry.

Food-based viral infection is another major health problem in human and animals. Currently, blueberry proanthocyanidins were tested against human norovirus growth in apple juice (AJ) and milk with 2% fat tannin-rich fraction from pomegranate rind (TFPR) inhibited the growth of human norovirus [18]. Hence, hydrolysable tannins are potential antiviral agents that used can be used in the food preservation to make food items more safe and preserve for prolonged period by using natural compounds.

Guava is a major tropical fruit which is also considered as a model system to study climacteric and non-climacteric fruit ripening process. Hence, it provides enough opportunities to understand post-harvest management of perishable fruits. In order to improve the shelf-life of fruits, various types of wax films, coating, and chemical treatments are used for long time. Tannins isolated from various natural sources act as preservatives due to their antibacterial and antioxidant properties. A coating material of tannic acid cross-linked with zein protein was used for coating on the guava fruit. Actually, zein is a prolamin (protein) isolated from aqueous alcohol-soluble fraction of corn (*Zea mays* L.) that earlier used to improve the shelf-life of guava fruit [19]. The coated guava fruits showed reduced ripening process and improved shelf-life compared with uncoated fruits. The coated fruits also showed more positive biochemical parameters associated with better fruit shelf-life, such as total soluble solids, respiration rate, and chlorophyll contents. Moreover, ethylene, ROS production, less water loss, and gas exchanges are also observed, which are attributed to cross-linking between tannic acid and zein [19]. Hence, tannin-based package and coating material may prove more effective and ecofriendly in food industry.

2.1.3 Functional foods or nutraceuticals industry

Normal food components mainly provide the energy and essential nutrients for the growth and development for animals including human, but food also consist of bioactive molecules or phytochemicals and their inclusion in the appropriate quantity can act as possible therapeutically active agents also known as nutraceuticals, for example, (poly) phenol-rich tannins [20]. Currently, study of molecular mechanisms and pathways such as cell proliferation, apoptosis, inflammation, differentiation, angiogenesis, DNA repair pathway, and carcinogens activation offer new therapeutic targets. The application of tannins has great potential as a nutraceuticals in order to prevent various diseases such as cancer, cardiovascular, kidney diseases, and diabetes. Major sources of tannins are fruits, vegetables, bark, wood, leaves, and seeds such as green tea, apples, cocoa, chocolate, grapes, apricots, and cherries. Among them, role of tannins present in the tea and coffee to prevent the cancers have been studied by large number of scientists. The green tea contains condensed tannin namely epigallocatechin gallate (EGCG) and epicatechin gallate (ECG), which have shown anticancerous activity in terms of inflammatory and antioxidant properties both in vivo and in vitro experimental systems [21]. Actually, EGCG decrease activation of NF-kB and (AP-1) TNF-α pathways simultaneously which ultimately reduced the production of IFN-γ. It can also enhance the apoptosis process through suppression of COX-2 enzyme that leads to the production of PGE2 (prostaglandins) in various types of cell lines of colon cancer cells, such as SW837, HT-29, and HCA-7 cells. Being a polyphenol, EGCG is a strong antioxidant that reduces the activity of nitric oxide (NO) and malondialdehyde (MDA) and conversely increases superoxide dismutase (SOD) activity in case of colonic mucosa. Hence, EGCG can improve the effect of cancer chemopreventive potency by preventing the cell proliferation, migration, and invasion of tumor in cancer patients [20].

In last two decades, a great interest has been emerged in the protective role of tannins against free radicals and reactive oxygen species produced inside cells, which caused degenerations and diseases such as cancer, atherosclerosis, and cardiovascular ailments. Proanthocyanidins have cardiovascular protective effect due to their antioxidant activity, inhibition of LDL oxidation, ability of vasodilation, antiplatelet activity, and protection against ischemia-reperfusion injury. Another tannin-based compound, gallic acid (3,4,5-trihydroxybenzoic acid), a naturally occurring with low molecular weight, plays very important in the protection of cardiovascular health through rejuvenating the antioxidant system which include large number of enzymes such as SOD, CAT, GPx, GRx, and GST which constitutes a scavenging system against the free radicals [22].

Diabetes mellitus, associated with high level of glucose concentration in the blood, is harmful for whole tissues in human body. Several investigations have shown that it can be reduced or managed by adding the appropriate amount of tannins in the nutrition or supplements of patients [23]. Because, tannins improve the glucose uptakes in body cell and simultaneously reduce the synthesis of adipocytes, hence act as the potential therapeutic agents. In a highly significant study, it shows that epigallocatechin gallate increases the glucose uptake by regulating insulin-signaling pathways, such as PI3K (phosphoinositide 3-kinase) and p38 MAPK (mitogen-activated protein kinase) activation and GLUT-4 translocation [24]. Tannins help reduction of blood glucose levels and offer antioxidants effects [23]. Therefore, it can be concluded that tannin-based foods are potential agents used as either nutraceutical or supplementary agents in food or medicine. A dimer of proanthocyanidin acts against the hyperglycemia that was created by sucrose feeding and by inhibiting the activity of α-glucosidase enzyme. The efficacy of proanthocyanidin has also been proved by molecular docking and strong inhibitory activity experiments.

It is already mentioned that nutraceuticals or functional foods have the health promoting effect on the human and animal health. Several epidemiological studies have clearly established a relationship between (poly) phenol-rich food items and human health. It has substantially enhanced consumer awareness about the tannin-rich diet, and their disease prevention capability; therefore, there is high demand of functional foods. A large number of tannin-based compounds are isolated and characterized from fruits and vegetables (**Table 1**). But, despite high cost incurred on extraction and separation of tannin, it offer only low yield which is a major cause of concern [25]. Additionally, it does not provide the pure content which hindered to test the efficacy and absorption of tannin-based foods in human subject. Grape-seed proanthocyanidins (GSP) were lyophilized to improve their in vivo absorbability. It was achieved by esterification of the water-soluble GSP and immobilized lipase. Lipophilicity was tested by 1-octanol/water partition coefficient as the absorbability parameter. Further, it was observed that GSP derivatives, 3′,5′-2-O-lauroyl epigallocatechin, 3′-O-lauroyl catechin, 3′-O-lauroyl epicatechin, and 3′,3″,5″-3-O-lauroyl epicatechin gallate show high level of radical scavenging activity; hence, it can be used as the strong antioxidant in food to prevent the major degenerative diseases and aging, which are generally caused by the free radicals in the tissues [25]. Recently, B-type proanthocyanidins were isolated and purified from fruits of elephant apple (*Dillenia indica* Linn.), their structural and bioactive properties were examined by using NMR, electron spray ionization, and matrix-assisted laser desorption ionization time of flight mass (MALDI-TOF) spectra. In this experiment, yield of 0.23% was achieved that is far greater than commercial grape-seed proanthocyanidins [26]. Hence, the EAPs may be used as promising functional food agents. In view of above health benefits provided by tannin-based compounds; they are now available in the dietary supplements. These contain biological extracts packed with both types of tannins and their consumption may provide health benefits. Recently, a concoction of ellagitannins, punicalagins, and polyphenols was commercialized by brand name, that is, Healing America Ellagitannin capsules and Ellagic Active Tablets.

2.2 Wood industry

Wood is the inseparable part of the furniture and several important industries. Wood contains organic acids, tannins, and lignocellulosic material which are most susceptible to biological, chemical, and physical decaying agents. Therefore, wood requires a large number of synthetic adhesives, glues, antitermite chemicals, and other coating materials in order to protect it. However, these materials have tremendously benefitted the wood industry, but they adversely affect the environment conditions. Because synthetic phenolics, amino resins, and formaldehyde used in wood industries are generally carcinogenic in nature. To overcome this problem, scientists are investigating natural materials of herbal or animal origins, such as tannins, that can be the best option or alternative material to be used in the wood industries [27].

In recent years, a lot of attempts have been made to improve the bio-durability and commercial properties of wood and wood-based furniture by using tannin-based preservatives. Although both condensed and hydrolysable tannins are used as adhesive, but mimosa tannin is proved to be the most effective wood glue, which attributed to good cross-linking, auto-condensation, poly condensations reactions, and hyper activity. Tannin-based adhesive is proved to be harder than pure synthetic adhesive due to great level of bonding with other aldehydes or different non aldehyde hardeners (glyoxal, furfuryl alcohol, hexamine, etc.) and lignocellulosic materials. Several industries in South Africa and America are using the mimosa and quebracho-based tannins that could reduce the formaldehyde-based emission from the industries. A similar technology has been used to produce the interior and exterior grade particle board largely used in the furniture industry [28]. Apart from this, catechin and gallic acid-based epoxide adhesives and starch-based adhesives from tannins were also produced. Epoxy adhesives are produced by reactions between catechin and epichloridrin via alkylation in the presence of unsaturated halogenated compound that leads to the oxidation. Tannin-based adhesives have been used for grinding wheels, angle grinder disks, and automotive brake pads matrices also.

Coating material is extremely useful to prevent wood surface from biotic and abiotic adversities like, rain, winter, and summer, and insects and microorganisms. Generally, coating material comprises of polyurethane and isocyanates. The urethane-based coating material is prepared by using the lignin and lignosulfonate/hydroxypropylate. Recently, Pinus tannins and di-isocyanates were used to prepare more effective coating material by exploiting hydroxypropylation and hydroxybutylation reactions which enhanced the bonding patterns between hydroxyl groups provided by tannins (flavonoids) and isocyanates [29]. However, tannins and isocyanate-based adhesives are good and environmental friendly, but these should be replaced with more bio-based material like tannins because they are naturally designed to protect wood against fungal attacks and natural decaying process of wood.

2.3 Medicine and pharmaceuticals

After the industrial revolution, large numbers of synthetic chemicals were used as drug molecules to treat numerous diseases but they left several adverse effects on the human and environment. Therefore, attention has been shifted to identify new alternative natural compounds that are to be clinically effective and create minimum adverse effects. A large number of natural compounds such as polyphenolic-based secondary metabolites, for example, tannins, are isolated and characterized as preventive therapeutic agents, which can be isolated from fruits, vegetables, or plants or expressed in the microorganism by metabolic engineering. Many studies have clearly shown that tannins are natural antioxidants linked with the prevention of degenerative diseases such as atherosclerosis, cardiovascular diseases, neurodegenerative diseases, and certain types of cancers by acting as antioxidants and antibacterial.

2.3.1 Tannins as preventive medicine

Reactive oxygen species such as, hydroxyl radical (HO•), superoxide anion ($O_2{\bullet}^-$), and peroxyl radical (ROO•) and the non-radicals like, hydrogen peroxide (H_2O_2) and hypochlorous acid (HOCl) are produced in biological systems. They adversely affect cellular protective systems which are responsible for many degenerative diseases in human. In order to overcome toxic effects of reactive oxygen species (ROS), tannins can be used as ROS scavenging agents. Actually, tannins have the ability to donate electron to a free radical or ROS and make them more stable compounds therefore, render less harmful effects on cellular environment [22, 30, 31]. Tannins also help by supporting antioxidant enzymes involved in the ROS scavenging activities, simultaneously inactivating the metal ions produced by free radicals. Many tannin-based products such as gallic acid (GA) (3,4,5-tri-hydroxybenzoic acid), proanthocyanidins, epigallocatechin gallate (EGCG), and ellagic acid-4-O-α-D-xylopyranoside have been tested and found highly effective as antioxidants. Gallic acid, isolated from many plant extracts, shows strong antioxidant properties responsible for the antioxidant and anticancer activities. Moreover, gallic acid derivatives (GADs) are present in large number of herbal medicines and formulations used for variety of diseases. Tannin derivatives like mucic acid gallate, mucic acid lactone gallate, monogalloylglucose, gallic acid, digalloylglucose, putranjivain A, galloyl-HHDP-glucose, elaeocarpusin, and chebulagic acid isolated from fruits of *Phyllanthus emblica* exhibit antioxidant activities that are already proved by the study of animals models. Currently, researchers have shifted their focus on the role of individual tannin molecules rather than group of compounds in the biological system for example, gallic acid, and epigallocatechin gallate (EGCG) from fruits and teas, respectively, are widely studied [22, 32].

2.3.2 Antibacterial properties of tannins

Synthetic antibiotics are being used as antibacterial agents for a long time in medical and animal sciences. But prolonged application of antibiotics lead to the development of resistance against the antimicrobial agents among the bacterial species attributed to selective evolutionary processes, a problem being faced by researcher world over. Nowadays, methicillin-resistant *Staphylococcus aureus* (MRSA), and multidrug-resistant pathogenic microorganisms are great health problems responsible for large number of morbidity and mortality in human population. Moreover, development of resistance to virtually all currently available antibiotics make situation more worsen. Therefore, it is of urgent need to discover new natural antimicrobial agents or antibiotics to cope with the development of antibiotics resistance [31].

However, many antibiotic resistance mechanisms are prevailed in the resilient microbial strains, but the mechanisms studied in *Staphylococcus aureus* RN4220 and IS-58 strains show that these particular strains have the capability to drain out the antibiotics from cytoplasm through proteinous membrane pumps. In a highly significant study, sub-concentration of pump inhibitors and tannins was used which significantly inhibited pump functions in both RN4220 and IS-58 strains [33].

Tannins have been used against the ATCC 43300 and MRSA clinical strains as membrane pump inhibitors and their mode of action was studied by using next-generation sequencing (NGS) in order to get deep understanding of antibacterial mechanisms at genome, transcriptome, and protein synthesis level. This investigation indicates that tannins mainly disrupt protein synthesis mechanisms by bringing major changes in ribosome pathways, which further caused a change in the translation processes in MRSA cells eventually leading to reduction in bacterial growth. Hence, tannins can be used as potential tools against the anti-MRSA agents in clinical application particularly, in antiseptic body solutions and antibacterial cream [31]. More recently, three ellagitannin-based tannins and isorugosin-A extracted in acetone from the fresh leaves of *Liquidambar formosana* showed high level of antibacterial activity against methicillin-resistant *Staphylococcus aureus* (MRSA) and *Pseudomonas aeruginosa* that attributed to tannins binding with membrane proteins, by polyphenolic acyl groups [30].

Tannin and its derivatives show great antibacterial properties which are used against a large number of bacterial species such as *Aeromanas*, *Bacillus*, *Clostridium*, *Enterobacter*, *Helicobacter*, *Klebsiella*, *Proteus*, *Pseudomonas*, *Shigella*, *Escherichia*, *Staphylococcus*, or *Streptococcus* and fungal species like *Aspergillus*, *Coniophora*, or *Penicillium*. Antibacterial activity of tannins particularly, polymeric proanthocyanidins proved highly effective against *Escherichia coli* and *Staphylococcus aureus*. It is attribute to binding of tannins with urinary tract epithelium and intestinal epithelium that prevent binding of disease causing organisms. In view of above findings, tannic acid is used as an inhibitor and immunomodulatory against multidrug resistant bacteria (MDR) [28, 33].

The antibacterial properties of tannins are not only studied in animals but in plants too. Several microorganisms cause the substantial loss in the fruit, vegetable, and plant species resulting in great economic loss. After the green revolution, huge amount of pesticides were used to prevent the bacterial and insect attacks in crops which lead to the environmental pollution and soil contamination. But bio-based pesticides or natural products can be the best option of chemical-based antibacterial agents. Recently, crude methanol extract of *Sapium baccatum* was used against the *Ralstonia solanacearum*, a causal agent of bacterial wilt of tomato. The extract mainly contains gallic acid, methyl gallate, corilagin, tercatain, chebulagic acid, chebulinic acid, and quercetin 3-O-α-L-arabinopyranoside which all show strong antibacterial activity except one tannin-based product, that is, quercetin 3-O-α-L-arabinopyranoside. In in vivo studies, the concentration of 2000 and 1000 μg/mL of crude extract reduced the development of tomato bacterial wilt by 83 and 63%, respectively [34].

2.3.3 Antifungal properties of tannins

The growth of fungi such as *Fusarium semitectum*, *F. fusiformis*, and *Alternaria altternata* can be hindered by gallic acid [22, 35]. The ethyl acetate extract and its sub-fraction from red raspberry (*Rubus idaeus*) fruit have high level of antifungal activities against the *Candida albicans*, *C. glabrata*, and *C. parapsilosis* strains of fungi due to their antimicrobial activities of tannins. These fungal strains form drug-resistant biofilms inside the oral cavity responsible for dental caries, periodontal disease, and denture stomatitis. However, the activity of extract was dose dependent, and 25 and 12.5 μg/mL of 80% ripe fruit extract was more effective as antiadherence or antibacterial agents against the microbial films formations [35].

2.3.4 Immunomodulatory activities of tannins

Immune system plays a very significant role to cope up with infectious agents like bacteria, virus, fungus, pollens, and parasites. Some experiments show that tannins modulate human immune system in a highly positive manner, thus tannins act as immunomodulatory agents in the battle against infectious diseases. Leishmaniosis, a disease that caused by parasitic protozoan's complex, comprise of more than 20 different species of *Leishmania* genus. Its conventional treatments are highly expensive, and lead to many side effects; moreover, protozoan resistance to treatments has been reported. Two most important tannins, gallic acid (GA) and ellagic acid (EA), were tested for antileishmania, cytotoxic, and immunomodulatory activities. Both GA and EA significantly reduced the infection and infectivity of macrophages infected by *L. major*. Moreover, both GA and EA induced high immunomodulatory activity of macrophage cells that proved by enhanced phagocytic capability, lysosomal volume, nitrite release, and intracellular calcium in macrophages. Therefore, tannins can be used as potential therapeutic agents against the leishmaniosis [36].

2.4 Role of tannins in animal husbandry

Bacterial and fungal infection is also a threat to the poultry, livestock, and animal husbandry which is responsible for high level of mortality. To overcome this problem, several antibiotics have been used for decades that proved to be very effective; consequently, it improves animal and poultry production in the world. But it is well known that extreme application of antibiotics promotes the antibiotic-resistant among the microorganisms in cattle [37–39]. Therefore, in-feed antibiotics and plant-based antibacterial agents, such as phytogenic compounds (e.g., tannin), have been discovered and promoted, which have great promises in future. In recent past, great attention has been given to antibacterial activities of tannins and their effects as dietary source in animal [37]. It has been concluded that tannins with saponins and essential oils can be used as in-feed antibiotics against bacteria, fungi, and yeasts. Because, tannins are toxic to bacteria and potentially inhibit growth of *Salmonella*, *Shigella*, *Staphylococcus*, *Pseudomonas*, and *Helicobacter pylori*, but it would be noteworthy that they show species specific antibacterial activity. Moreover, tannin-containing forage in cattle diets helps to control animal pasture bloating, intestinal parasite, and disease causing bacteria in ruins of animals. Tannins can hinder microbial growth by using several mechanisms including lack of nutrient to bacterial cell, inactivate vital extracellular enzymes, inhibition of oxidative phosphorylation, chelation of metal ions, and complex formation with membrane and proteins [38]. It has been seen that condensed type of tannins are mainly present in forage legumes, trees, shrubs, tree leaves, and browse shrubs, but their concentration vary from species to species that influenced by environmental conditions also [40]. Tannins from mimosa (HT), chestnut (HT), and quebracho (CT) have been used as in-feed antibiotics in animals [41]. But the major challenge for tannins as antibiotics is the lack of systematic and comprehensive studies on the various aspects such as doses, side effects on digestions simultaneously prolong use can develop resistance against the in-feed antibiotics as in case of normal antibiotics [40]. Moreover, tannins are antinutrient factors for monogastric animals and poultry.

Tannins can also act as the antinutrients in rumens of livestock due to their binding to vital biomolecules in biological systems. Several adverse effects such as availability of nutrients, metal ions chelation, binding with proteins and hinder the growth of beneficial microflora have been observed in the cattle gut. To test the adverse effect of tannin as diet component on lamb gut microflora and fermentation was studied. Both types of tannins, that is, hydrolysable and condensed with 4% extract of chestnut (*Castanea sativa*, *Caesalpinia spinosa*), *mimosa* (*Acacia negra*), and *gambier* (*Uncaria gambir*) feed to lamb. The results show that tannins meagerly affect gut microflora including fungi in the lamb gut for 45 days. Simultaneously, it also shows that high level of tannin inclusion in diet proved as antimicrobial agent against the harmful methanogens and protozoa without affecting ruminal fermentation [42].

3. Nanotechnology and tannin

Cancer is a fatal disease and its occurrence in the human population is the major cause of concern. However, the role of tannins as chemopreservants in the cure of cancer has been widely discussed by many researchers [42]. The chemoprevention "is a means of cancer management by which the occurrence of the disease can be entirely prevented, slowed, or reversed via administration of one or more naturally occurring and/or synthetic compounds" [43]. Currently, target-based delivery of anticancer agents to the site of cancer or tumor is major challenge. In order to target tumor at nanolevel, cancer nanotechnology has made tremendous progress in last one decade. It is assumed that, if drug is delivered at nanolevel at site of tumor with a high level of specificity so that cancer can be better managed. Nanoparticles of various tannin-based compounds are also prepared but their toxicity to normal body cells left major side effects. Therefore, encapsulation of many types of tannin like, epigallocatechin-3-gallate in chitosan-tripolyphosphate nanoparticles was investigated for target-based delivery to tumor. It is well known that (−)-epigallocatechin-3-gallate, (−)-epigallocatechin, (−)-epicatechin-3-gallate and (−)-epicatechin act as anticancer and antioxidant agents. The nanoemulsions and liposomes of tannins have proven highly effective in target-based delivery of anticancer drugs in case of HepG2 cells [44]. So far, nanoencapsulation method is proved only in in vitro studies and animal models but it is rarely proved effective in normal and cancerous cells.

3.1 Antiviral activity of tannin

In current era, viral infection is the major threat to the human and animal population. Tannins also show antiviral activity in case of several diseases such as HIV, herpes simplex virus 2 (HSV-2). In case of herpes simplex virus 2 silver nanoparticles with tannic acid (TA-AgNPs) act as microbicide by preventing adsorption of viral particle in the body. Additionally, tannins also provide the better adjutant properties for example, substantial improvement of production of IFN-gamma+ CD8+ T-cells, activated B cells, and plasma cells. In case of spleen also, tannins promotes production of higher amount of IFN-gamma+ NK cells and effector-memory CD8+ T-cells particularly, in case of second challenge against HSV-2 immune response [45].

Free-living protozoa species of *Acanthamoeba* genus generally cause significant infections of keratitis and encephalitis in human. Acanthamoeba keratitis is a cornea related infection that adversely affected eye vision. It is resistance to current available therapy. To overcome this problem, both pure silver and gold nanoparticle and tannic acid-modified of nanoparticles of silver and gold were prepared and their activities were tested against the clinical strains of *Acanthamoeba* spp. The tannic acid-modified nanoparticles proved more effective and less toxic to eye infection. Moreover, tannic acid-modified silver nanoparticles were well absorbed by the trophozoites eventually inhibits germination of cyst which is a major stage of life cycle of amoebae parasite [46]. So that it can be concluded that tannin-based nanoparticles are more effective than pure silver and gold particle.

4. Leather industry

In leather industry, tannins are generally used to convert animal hide into leather. Here, the main role of tannins is to protect leather from microorganism and heat related deterioration [47]. Tanning industry is thought to be the oldest industry and was started in north western regions of Europe after the Roman conquest [48]. Tannins bind with the skin proteins and protect it from petrification, which is owed to the antibacterial property of tannins. This is due to the chemical bonds established between collagen, the main constitutive protein of skin, for example, collagens, and the tannins present in the vegetable materials. It is estimated that about 15–45% tannins binds with collagens per dry weight. After the industrial revolution, chromium-based tanning was introduced to achieve fast and speedy leather production. But chromium is a potentially carcinogenic and creates high level of pollution and soil contamination. Additionally, it limits recovery and reuse of wastes from leather industries. Water from leather industry creates more pollution, and increase biological oxygen demand (BOD) and chemical oxygen demand (COD) in polluted water. In order to overcome this problem, plant-based tannins can be substitute for chromium in leather processing. It is already mentioned that vegetable tannins were used in the leather industry since historical times. But now, tannins from different parts of plants have been utilized in leather productions [49]. Plant tannins offer many benefits such as high quality and thermal stable leather products.

5. Other industrial applications

In addition to above mentioned applications, tannins are also used in diverse types of industries, such as paper industry where high pressure mimosa tannin impregnated alpha cellulose paper is prepared. This impregnated paper offer more abrasive resistance, adhesion properties, water vapor resistance, and staining properties [50]. Recently, natural tannin-based foam without any formaldehyde is prepared that used as acoustic absorbers, metal ion adsorption, panels crash protection, packaging, etc., but low mechanical strength of tannin foams is major impediment in its further applications [51].

Beverage industry is well known for tannins applications particularly, in case of the wine making. Like leather industry, wine making is also very old industry since historic time. Tannins are used in wine to provide color formations, antioxidants, aroma, proteins precipitations, and flavor development. The source of tannins in the wine is grapes skin, seeds, and addition of oak flakes which add market values to wine [52].

6. Future prospectus

Tannins are phenolic-based secondary metabolites that are present in the plant kingdom, including algae. Actually, tannins produced in the plant body and involved in the plant protections and act as antimicrobial, antiparasitic, anthelmintic, antiviral, antioxidant, and deferred cattle. Hence, they help plants to fight various types' infections. In addition to biological roles, they also play very important roles in industrial sector, animal feeding, mining, chemical industry, and tanning industry. But there are several limitations associated with the tannins. The main negative effect of tannins as food is their absorptions and binding with various types of biomolecules such as proteins, starch, and metal ions in the digestive system, hence hinders their nutritional availability to human and animals, for example, proanthocyanidins. Some experiments show that complex of tannins and proteins are resistant to various types of proteases in animal digestive system that make proteins unavailable for livestock nutrition. Dietary tannins bind with the proline-rich proteins and as result two types of soluble and insoluble complexes formed which is responsible for astringent sensation [53]. The astringency feeling is perceived by the tongue in the form of diffuse feeling associated with extreme dryness and roughness in mouth [54]. Some experiments also show that tannins also decrease activities of intestinal microflora, consequently less absorption of organic matter and soluble fiber that attributed to damage the mucosal lining of the digestive system. Moreover, high dose of tannins like catechin used in supplements can cause renal failure, hepatitis, fever, hemolytic anemia, thrombocytopenia, and skin disorders. Due to the structural complexities, tannins are also considered as potential pollutants in industries, where tannins are used as the major substrate.

Single-meal bioavailability studies have shown major antinutrient activity of tannins is metal ions chelation that cause severe deficiency of essential minerals in human. In this series iron deficiency is most prevalent in population particularly in the developing countries. Recently, single-meal studies in case of hydrolysable and oligomeric catechin and epicatechin tannins (tea and tannic acid) have conducted. It shows reduced iron bioavailability in diet particularly in long term, but it needs more investigation by using appropriate animal model systems in order to study antinutritional role of tannins [55]. Moreover, tannins not only affect iron availability, but also iron metabolism; the ferritin, an iron transport protein, is adversely affected by tannin binding in soybean seed ferritin (SSF) and consequently changes the tertiary/quaternary structure of the protein.

In view of current scientific investigation, it has now become possible that tannins can be exploited in a better way because they are major sustainable raw materials for green chemistry in future. Recently, several tannase or tannin acyl hydrolase enzymes have isolated, characterized, and classified from microorganism and fungi, hence toxic effect of tannins is reduced through hydrolysis and oxidation [56, 57]. Various species of filamentous fungi that produced tannase are used for bioremediation in leather tannin industries.

Several studies have been conducted to overcome the adverse effect of tannins in the food item like, fruit and vegetable. In a major breakthrough, lactic acid bacteria fermentation-like incubation is exploited in *Xuan Mugua* fruits, as a result up to 70% tannin content reduced with substantially reduced astringency, hence that method can be used in the food industry [58].

7. Conclusions

Tannins are phenolic-based secondary metabolites that are present in the plant kingdom, including algae. Actually, tannins produced in the plant body involved in the plant protections and act as antimicrobial, antiparasitic, anthelmintic, antiviral, antioxidant, and deferred cattle. Hence, they help plants to fight various types of infections. In addition to biological roles, they also play very important roles in industrial sector, animal feeding, mining, chemical industry, and tanning industry. But there are several limitations associated with the tannins applications. The main negative effect of tannins as food are their absorptions and binding with various types of biomolecules such as, proteins, starch, and metal ions in the digestive system; hence, they hinder their nutritional availability to human and animals.

Acknowledgements

I would like to thank to my Institution, GGD SD College (Panjab University) India, for providing me all the facilities and support for preparation of this manuscript.

Conflict of interest

Authors declare no conflict of interest.

Notes/Thanks/Other declarations

Authors also acknowledge the contribution of publications which could not be included in the reference section due to lack of space.

Acronyms and abbreviations

PA	proanthocyanidins
LDPE	polyethylene and low-density polyethylene
LLDPE	linear low-density polyethylene
WVP	water vapor permeability
SDF	soluble dietary fiber
MRSA	methicillin-resistant *Staphylococcus aureus*
TFPR	tannin-rich fraction from pomegranate rind
EGCG	epigallocatechin gallate
MALDI-TOF	electron spray ionization and matrix-assisted laser desorption ionization time of flight mass
TA-AgNPs	silver nanoparticles with tannic acid
HSV-2	herpes simplex virus 2

Author details

Akhlash P. Singh[1]* and Sunil Kumar[2]

1 Department of Biochemistry, GGD SD College, Panjab University, Chandigarh, India

2 ICAR-Central Institute of Post-Harvest Engineering and Technology (ICAR-CIPHET), Abohar, Punjab, India

*Address all correspondence to: akhlash@gmail.com; akhlash@ggdsd.ac.in

References

[1] Catarino M, Silva A, Cardoso S. Fucaceae: A source of bioactive phlorotannins. International Journal of Molecular Sciences. 2017;**18**(6):1327

[2] Choubey S, Goyal S, Varughese LR, Kumar V. Probing gallic acid for its broad spectrum applications. Mini Reviews in Medicinal Chemistry. 2018;**18**(15):1283-1293

[3] Sanjeewa KKA, Kim EA, Son KT, Jeon YJ. Bioactive properties and potentials cosmeceutical applications of phlorotannins isolated from brown seaweeds: A review. Journal of Photochemistry and Photobiology B: Biology. 2016;**162**:100-105

[4] Horvath PJ. The nutritional and ecological significance of tannins and related polyphenols [M.S. thesis]. Ithaca, New York, USA : Cornell University; 1981

[5] Serrano J, Puupponen-Pimiä R, Dauer A, Aura AM, Saura-Calixto F. Tannins: Current knowledge of food sources, intake, bioavailability and biological effects. Molecular Nutrition & Food Research. 2009;**53**(S2):S310-S329

[6] Koleckar V, Kubikova K, Rehakova Z, Kuca K, Jun D, Jahodar L, et al. Condensed and hydrolysable tannins as antioxidants influencing the health. Mini Reviews in Medicinal Chemistry. 2008;**8**(5):436-447

[7] Porter LJ, Flavonoids T. In: Harborne JB, editor. Advances in Research since 1986. London: Chapman & Hall; 1986. pp. 23-55

[8] Khanbabaee K, Van Ree T. Tannins: Classification and definition. Natural Product Reports. 2001;**18**:641_9

[9] https://www.grandviewresearch.com/industry-analysis/tannin-market. Apr 2017

[10] Luckeneder P, Gavino J, Kuchernig R, Petutschnigg A, Tondi G. Sustainable phenolic fractions as basis for furfuryl alcohol-based co-polymers and their use as wood adhesives. Polymers. 2016;**8**:396

[11] Missio AL, Mattos BD, Ferreira DDF, Magalhães WL, Bertuol DA, Gatto DA, et al. Nanocellulose-tannin films: From trees to sustainable active packaging. Journal of Cleaner Production. 2018;**184**: 143-151

[12] Kamari A, Phillip E. Chitosan, gelatin and methylcellulose films incorporated with tannic acid for food packaging. International Journal of Biological Macromolecules. 2018;**120**:1119-1126

[13] Liu R, Cong X, Song Y, Wu T, Zhang M. Edible gum–phenolic–lipid incorporated gluten films for food packaging. Journal of Food Science. 2018;**83**(6):1622-1630

[14] Wang Y, Li J, Li B. Nature-inspired one-step green procedure for enhancing the antibacterial and antioxidant behavior of a chitin film: Controlled interfacial assembly of tannic acid onto a chitin film. Journal of Agricultural and Food Chemistry. 2016;**64**(28):5736-5741

[15] Song GB, Xu J, Zheng H, Feng Y, Zhang WW, Li K, et al. Novel soluble dietary fiber–tannin self-assembled film: A promising protein protective material. Journal of Agricultural and Food Chemistry. 2015;**63**(24):5813-5820

[16] Hintz T, Matthews KK, Di R. The use of plant antimicrobial compounds for food preservation. BioMed Research International. 2015. 12 p. Article ID: 246264

[17] Lipińska L, Klewicka E, Sójka M. Structure, occurrence and biological activity of ellagitannins:

A general review. Acta Tech. 2014;**13**(3):289-299

[18] Yang Q, Wang L, Gao J, Liu X, Feng Y, Wu Q, et al. Tannin-rich fraction from pomegranate rind inhibits quorum sensing in *Chromobacterium violaceum* and biofilm formation in *Escherichia coli*. Foodborne Pathogens and Disease. 2016;**13**(1):28-35

[19] Santos TM, Men de SáFilho MS, Silva EDO, da Silveira MR, de Miranda MRA, Lopes MM, et al. Enhancing storage stability of guava with tannic acid-crosslinked zein coatings. Food Chemistry. 2018;**257**:252-258

[20] Rescigno T, Tecce MF, Capasso A. Protective and restorative effects of nutrients and phytochemicals. The Open Biochemistry Journal. 2018;**12**:46

[21] Smeriglio A, Barreca D, Bellocco E, Trombetta D. Proanthocyanidins and hydrolysable tannins: Occurrence, dietary intake and pharmacological effects. British Journal of Pharmacology. 2017;**174**(11):1244-1262

[22] Badhani B, Sharma N, Kakkar R. Gallic acid: A versatile antioxidant with promising therapeutic and industrial applications. RSC Advances. 2015;**5**(35):27540-27557

[23] Ajebli M, Eddouks M. The promising role of plant tannins as bioactive antidiabetic agents. Current Medicinal Chemistry. 2018;**25**:1

[24] Anderson RA, Polansky MM. Tea enhances insulin activity. Journal of Agricultural and Food Chemistry. 2002;**50**:7182-7186

[25] Chen M, Yu S. Lipophilized grape seed proanthocyanidin derivatives as novel antioxidants. Journal of Agricultural and Food Chemistry. 2017;**65**(8):1598-1605

[26] Fu C, Yang D, Peh WYE, Lai S, Feng X, Yang H. Structure and antioxidant activities of proanthocyanidins from elephant apple (*Dillenia indica* Linn.). Journal of Food Science. 2015;**80**(10):C2191-C2199

[27] Hubbe MA, Pizzi A, Zhang H, Halis R. Critical links governing performance of self-binding and natural binders for hot-pressed reconstituted lignocellulosic board without added formaldehyde: A review. BioResources. 2017;**13**:2049-2115

[28] Pizzi A, Stracke P, Trosa A. Industrial tannin/hexamine low-emission exterior particleboards. European Journal of Wood and Wood Products. 1997;**55**:168-168

[29] García DE, Glasser WG, Pizzi A, Paczkowski SP, Laborie MP. Modificationof condensed tannins: From polyphenol chemistry to materials engineering. New Journal of Chemistry. 2016;**40**:36-49

[30] Obiang-Obounou BW, Ryu GH. The effect of feed moisture and temperature on tannin content, antioxidant and antimicrobial activities of extruded chestnuts. Food Chemistry. 2013;**141**(4):4166-4170

[31] Siti-Noor-Adnalizawati Adnan NI, Yaacob WA. Disruption of methicillin-resistant *Staphylococcus aureus* protein synthesis by tannins. Germs. 2017;**7**(4):186

[32] Yang B, Liu P. Composition and biological activities of hydrolyzable tannins of fruits of *Phyllanthus emblica*. Journal of Agricultural and Food Chemistry. 2014;**62**(3):529-541

[33] Tintino SR, Morais-Tintino CD, Campina FF, Costa MDS, Menezes IR, de Matos YML, et al. Tannic acid affects the phenotype of *Staphylococcus aureus* resistant to tetracycline and erythromycin

by inhibition of efflux pumps. Bioorganic Chemistry. 2017;**74**:197-200

[34] Vu TT, Kim H, Tran VK, Vu HD, Hoang TX, Han JW, et al. Antibacterial activity of tannins isolated from *Sapium baccatum* extract and use for control of tomato bacterial wilt. PLoS One. 2017;**12**(7):e0181499

[35] Dutreix L, Bernard C, Juin C, Imbert C, Girardot M. Do raspberry extracts and fractions have antifungal or anti-adherent potential against *Candida* spp.? International Journal of Antimicrobial Agents. 2018;**52**(6):947-953

[36] de Moraes Alves MM, Brito LM, Souza AC, Queiroz BCSH, de Carvalho TP, Batista JF, et al. Gallic and ellagic acids: Two natural immunomodulator compounds solve infection of macrophages by *Leishmania major*. Naunyn-Schmiedeberg's Archives of Pharmacology. 2017;**390**(9):893-903

[37] Redondo LM, Chacana PA, Dominguez JE, Fernandez Miyakawa MED. Perspectives in the use of tannins as alternative to antimicrobial growth promoter factors in poultry. Frontiers in Microbiology. 2014;**5**:118

[38] Liu XL, Hao YQ, Jin L, Xu ZJ, McAllister TA, Wang Y. Anti-*Escherichia coli* O157: H7 properties of purple prairie clover and sainfoin condensed tannins. Molecules. 2013;**18**:2183-2199

[39] Chattopadhyay MK. Use of antibiotics as feed additives: A burning question. Frontiers in Microbiology. 2014;**5**:33

[40] Huang Q, Hu T, Xu Z, Jin L, McAllister TA, Acharya S, et al. 286 structural composition and protein precipitation capacity of condensed tannins from purple prairie clover (*Dalea purpurea* Vent.). Journal of Animal Science. 2017;**95**(suppl_4):141-142

[41] Min BR, Hernandez K, Pinchak WE, Anderson RC, Miller JE, Valencia E. Effects of plant tannin extracts supplementation on animal performance and gastrointestinal parasites infestation in steers grazing winter wheat. Open Journal of Animal Sciences. 2015;**5**(03):343

[42] Salami SA, Valenti B, Bella M, O'Grady MN, Luciano G, Kerry JP, et al. Characterisation of the ruminal fermentation and microbiome in lambs supplemented with hydrolysable and condensed tannins. FEMS Microbiology Ecology. 2018;**94**(5):fiy061

[43] Santos IS, Ponte BM, Boonme P, Silva AM, Souto EB. Nanoencapsulation of polyphenols for protective effect against colon–rectal cancer. Biotechnology Advances. 2013;**31**(5):514-515

[44] Donsì F, Sessa M, Mediouni H, Mgaidi A, Ferrari G. Encapsulation of bioactive compounds innanoemulsion-based delivery systems. Procedia Food Science. 2011;**1**:1666-1671

[45] Orłowski P, Kowalczyk A, Tomaszewska E, Ranoszek-Soliwoda K, Węgrzyn A, Grzesiak J, et al. Antiviral activity of tannic acid modified silver nanoparticles: Potential to activate immune response in Herpes Genitalis. Viruses. 2018;**10**(10):524

[46] Padzik M, Hendiger EB, Chomicz L, Grodzik M, Szmidt M, Grobelny J, et al. Tannic acid-modified silver nanoparticles as a novel therapeutic agent against *Acanthamoeba*. Parasitology Research. 2018;**117**(11):3519-3525

[47] Kaygusuz M, Meyer M, Junghans F, Aslan A. Modification of leather surface with atmospheric pressure plasma and nanofinishing. Polymer-Plastics Technology and Engineering. 2018;**57**:260-268

[48] Van Driel-Murray C. Tanning and leather. In: Oleson JP, editor. The Oxford Handbook of Engineering and Technology in the Classical World. New York, NY, USA: Oxford University Press; 2009. pp. 483-495

[49] Falcão L, Araújo M. Vegetable tannins used in the manufacture of historic leathers. Molecules (Basel, Switzerland). 2018;**23**(5):1081. DOI: 10.3390/molecules23051081

[50] Abdullah UHB, Pizzi A, Zhou X. High pressure paper laminates from mimosa tannin resin. International Wood Products Journal. 2014;**5**:224-227

[51] Delgado-Sánchez C, Letellier M, Fierro V, Chapuis H, Gérardin C, Pizzi A, et al. Hydrophobisation of tannin-based foams by covalent grafting of silanes. Industrial Crops and Products. 2016;**92**:116-126

[52] Harbertson JF, Parpinello GP, Heymann H, Downey MO. Impact of exogenous tannin additions on wine chemistry and wine sensory character. Food Chemistry. 2012;**131**:999-1008

[53] He M, Tian H, Luo X, Qi X, Chen X. Molecular progress in research on fruit astringency. Molecules. 2015;**20**(1):1434-1451

[54] Kaspchak E, Mafra LI, Mafra MR. Effect of heating and ionic strength on the interaction of bovine serum albumin and the antinutrients tannic and phytic acids, and its influence on in vitro protein digestibility. Food Chemistry. 2018;**252**:1-8

[55] Delimont NM, Haub MD, Lindshield BL. The impact of tannin consumption on iron bioavailability and status: A narrative review. Current Developments in Nutrition. 2017;**1**(2):1-12

[56] de las Rivas B, Rodríguez H, Anguita J, Muñoz R. Bacterial tannases: Classification and biochemical properties. Applied Microbiology and Biotechnology. 2018;**103**(2):1-21

[57] Prigione V, Spina F, Tigini V, Giovando S, Varese GC. Biotransformation of industrial tannins by filamentous fungi. Applied Microbiology and Biotechnology. 2018;**102**(24):10361-10375

[58] Shang YF, Cao H, Ma YL, Zhang C, Ma F, Wang CX, et al. Effect of lactic acid bacteria fermentation on tannins removal in *Xuan Mugua* fruits. Food Chemistry. 2019;**274**:118-122

[59] Sieniawska E, Baj T. Tannins. In: Pharmacognosy. Fundamentals, Applications and Strategies. Academic Press; 2017. pp. 199-232. DOI: doi.org/10.1016/B978-0-12-802104-0.00010-X